Pasture-Associated Poisoning in Grazing Animals

Guest Editor

Dominique-Marie Votion

MDPI

Basel • Beijing • Wuhan • Barcelona • Belgrade • Novi Sad • Cluj • Manchester

Guest Editor
Dominique-Marie Votion
Department of Functional
Sciences
University of Liège
Liege
Belgium

Editorial Office
MDPI AG
Grosspeteranlage 5
4052 Basel, Switzerland

This is a reprint of the Special Issue, published open access by the journal *Animals* (ISSN 2076-2615), freely accessible at: https://www.mdpi.com/journal/animals/special_issues/5PM3UB86C8.

For citation purposes, cite each article independently as indicated on the article page online and as indicated below:

Lastname, A.A.; Lastname, B.B. Article Title. *Journal Name* **Year**, *Volume Number*, Page Range.

ISBN 978-3-7258-6728-8 (Hbk)
ISBN 978-3-7258-6729-5 (PDF)
https://doi.org/10.3390/books978-3-7258-6729-5

Cover image courtesy of Dominique-M. Votion

Contents

About the Editor

Dominique-Marie Votion

Dominique-M. Votion is a Professor at the Faculty of Veterinary Medicine, University of Liège, where she teaches veterinary pharmacology and toxicology, and she serves as Vice Director of the Fundamental and Applied Research for Animals & Health (FARAH) Research Unit. Her research interests focus on preventive strategies in equine health, with particular emphasis on pasture-associated poisoning and muscle disorders. Over the past two decades, her work has centered on atypical myopathy, an environmental intoxication affecting grazing equids and known in the United States as seasonal pasture myopathy, which is linked to ingestion of toxins associated with certain maple trees and is of increasing concern in Europe due to climate change and the invasive nature of the implicated species. Her research approach has combined epidemiology, histopathology, specialized biochemistry, and, more recently, omics technologies, contributing to the identification of the causative toxins and a better understanding of the disease pathophysiology. Beyond atypical myopathy, these methodologies have broader applications in the prevention of pasture-related intoxications and other equine health challenges. Her achievements include the development of innovative diagnostic tools for muscle disorders, such as acylcarnitine profiling and high-resolution respirometry for the assessment of mitochondrial function, as well as research programs aimed at addressing emerging environmental toxicities. In 2004, she founded the Atypical Myopathy Alert Group (AMAG), an informal European epidemiological surveillance network bringing together horse owners, equine practitioners, national surveillance systems, and academic institutions to issue alerts and share information on atypical myopathy.

animals

MDPI

Editorial

Happiness Is in the Field

Dominique-Marie Votion

Department of Functional Sciences, Faculty of Veterinary Medicine, Pharmacology and Toxicology, Fundamental and Applied Research for Animals & Health (FARAH), University of Liège, 4000 Liège, Belgium; dominique.votion@uliege.be

Editorial

Common sense suggests that 'happiness is in the field', as grasslands are considered environments that promote the health and well-being of grazing animals. However, they can present toxicological risks that are sometimes poorly known, so there is certainly a need to update our knowledge of pasture-associated poisoning.

This Special Issue gathers articles that explore both well-documented and emerging threats to livestock and horses linked to their grazing environment.

The collection includes case reports and investigations on acute poisonings caused by well-known toxic plants such as oak (*Quercus* spp.), where ingestion of large quantities of acorns can lead to death in both cattle [1] and horses [2,3]. Another study highlights the role of cyanogenic plants like sorghum, in which environmental stressors such as drought may exacerbate the production of toxins like dhurrin, triggering cyanide poisoning in cattle [4].

One of the most emblematic examples of a growing pasture-associated disease is equine atypical myopathy [5]. This condition, linked to ingestion of protoxins from seeds or seedlings of the sycamore maple (*Acer pseudoplatanus*), has emerged over the last two decades as a major cause of death in pastured horses across Europe. Although *A. pseudoplatanus* is considered an indigenous tree, its role in pasture toxicity was only recognised a decade ago [6]. Moreover, this syndrome is no longer restricted to horses as *A. pseudoplatanus poisoning* has been confirmed in other species [7–10]. This illustrates that pasture poisoning is not a static phenomenon: it is shaped by changing ecosystems, grazing behaviour [11], and climate changes. Insights into the equine faecal microbiota in the context of atypical myopathy suggest that gut microbial activity may modulate toxin metabolism and susceptibility [12]. Understanding these host–toxin interactions could open up new approaches to prevention.

Environmental toxicology also encompasses chronic and subclinical exposures. One study in this issue assesses the concentrations of toxic and essential trace elements in cow's milk in north-eastern Brazil, showing that proximity to major roads correlates with high levels of lead in milk, a potential public health problem [13]. Such findings remind us that contamination does not always present as acute disease but can emerge insidiously through bioaccumulation and long-term exposure.

Throughout history, toxic threats have evolved, from classic cases involving heavy metals [14,15] to current concerns such as pesticides [16] and per- and polyfluoroalkyl substances (PFAS) [17]. Effective management of pasture-associated poisoning requires both robust diagnostic tools [18] and a mechanistic understanding of toxic processes [19], enabling not just treatment but prevention [20].

As the welfare of grazing animals is intimately linked to pasture [21–23], we need to ensure that pasture remains a safe environment, whether for leisure animals or for livestock in food production systems. This Special Issue reminds us of the necessity to be aware of

https://doi.org/10.3390/ani15213087

the risks associated with pasture to strengthen our ability to protect animal health in an ever-changing environment.

Funding: This research received no external funding.

Conflicts of Interest: The authors declare no conflict of interest.

References

1. Eppe, J.; Bayrou, C.; Casalta, H.; Cassart, D.; Gille, L.; Stipulanti, M.; Versyp, J.; Sartelet, A. Oak Acorn Poisoning in Cattle during Autumn 2022: A Case Series and Review of the Current Knowledge. *Animals* **2023**, *13*, 2678. [CrossRef]
2. Smith, S.; Naylor, R.J.; Knowles, E.J.; Mair, T.S.; Cahalan, S.D.; Fews, D.; Dunkel, B. Suspected acorn toxicity in nine horses. *Equine Vet. J.* **2015**, *47*, 568–572. [CrossRef]
3. Hermange, T.; Ruault, B.; Courouсé, A. Retrospective Study of 25 Cases of Acorn Intoxication Colitis in Horses between 2011 and 2018 and Factors Associated with Non-Survival. *Animals* **2024**, *14*, 599. [CrossRef] [PubMed]
4. Giantin, S.; Franzin, A.; Brusa, F.; Montemurro, V.; Bozzetta, E.; Caprai, E.; Fedrizzi, G.; Girolami, F.; Nebbia, C. Overview of Cyanide Poisoning in Cattle from *Sorghum halepense* and *S. bicolor* Cultivars in Northwest Italy. *Animals* **2024**, *14*, 743. [CrossRef] [PubMed]
5. François, A.C.; Renaud, B.; Kruse, C.J.; Marcillaud-Pitel, C.; Votion, D.M. An ongoing concern: 20 years of research on equine atypical myopathy. *Vet. Rec.* **2024**, *195*, e5039. [CrossRef]
6. Votion, D.M.; van Galen, G.; Sweetman, L.; Boemer, F.; de Tullio, P.; Dopagne, C.; Lefere, L.; Mouithys-Mickalad, A.; Patarin, F.; Rouxhet, S.; et al. Identification of methylenecyclopropyl acetic acid in serum of European horses with atypical myopathy. *Equine Vet. J.* **2014**, *46*, 146–149. [CrossRef] [PubMed]
7. Renaud, B.; Kruse, C.J.; François, A.C.; Grund, L.; Bunert, C.; Brisson, L.; Boemer, F.; Gault, G.; Ghislain, B.; Petitjean, T.; et al. *Acer pseudoplatanus*: A Potential Risk of Poisoning for Several Herbivore Species. *Toxins* **2022**, *14*, 512. [CrossRef]
8. Hirz, M.; Gregersen, H.A.; Sander, J.; Votion, D.M.; Schanzer, A.; Kohler, K.; Herden, C. Atypical myopathy in 2 Bactrian camels. *J. Vet. Diagn. Investig.* **2021**, *33*, 961–965. [CrossRef]
9. Bunert, C.; Langer, S.; Votion, D.M.; Boemer, F.; Muller, A.; Ternes, K.; Liesegang, A. Atypical myopathy in Pere David's deer (Elaphurus davidianus) associated with ingestion of hypoglycin A. *J. Anim. Sci.* **2018**, *96*, 3537–3547. [CrossRef]
10. Bochnia, M.; Ziemssen, E.; Sander, J.; Stief, B.; Zeyner, A. Methylenecyclopropylglycine and hypoglycin A intoxication in three Pere David's Deers (Elaphurus davidianus) with atypical myopathy. *Vet. Med. Sci.* **2020**, *7*, 998–1005. [CrossRef]
11. Aboling, S. Do Poisonous Plants in Pastures Communicate Their Toxicity? Meta-Study and Evaluation of Poisoning Cases in Central Europe. *Animals* **2023**, *13*, 3795. [CrossRef]
12. François, A.C.; Cesarini, C.; Taminiau, B.; Renaud, B.; Kruse, C.J.; Boemer, F.; van Loon, G.; Palmers, K.; Daube, G.; Wouters, C.P.; et al. Unravelling Faecal Microbiota Variations in Equine Atypical Myopathy: Correlation with Blood Markers and Contribution of Microbiome. *Animals* **2025**, *15*, 354. [CrossRef] [PubMed]
13. Oliveira Filho, E.F.; López-Alonso, M.; Vieira Marcolino, G.; Castro Soares, P.; Herrero-Latorre, C.; Lopes de Mendonça, C.; de Azevedo Costa, N.; Miranda, M. Factors Affecting Toxic and Essential Trace Element Concentrations in Cow's Milk Produced in the State of Pernambuco, Brazil. *Animals* **2023**, *13*, 2465. [CrossRef]
14. Angon, P.B.; Islam, M.S.; Kc, S.; Das, A.; Anjum, N.; Poudel, A.; Suchi, S.A. Sources, effects and present perspectives of heavy metals contamination: Soil, plants and human food chain. *Heliyon* **2024**, *10*, e28357. [CrossRef]
15. Jomova, K.; Alomar, S.Y.; Nepovimova, E.; Kuca, K.; Valko, M. Heavy metals: Toxicity and human health effects. *Arch. Toxicol.* **2025**, *99*, 153–209. [CrossRef]
16. Garud, A.; Pawar, S.; Patil, M.S.; Kale, S.R.; Patil, S. A Scientific Review of Pesticides: Classification, Toxicity, Health Effects, Sustainability, and Environmental Impact. *Cureus* **2024**, *16*, e67945. [CrossRef] [PubMed]
17. Bonato, T.; Pal, T.; Benna, C.; Di Maria, F. Contamination of the terrestrial food chain by per- and polyfluoroalkyl substances (PFAS) and related human health risks: A systematic review. *Sci. Total Environ.* **2025**, *961*, 178337. [CrossRef]
18. Draghi, S.; Fehri, N.E.; Ateş, F.; Özsobacı, N.P.; Tarhan, D.; Bilgiç, B.; Dokuzeylül, B.; Yaramış, Ç.P.; Ercan, A.M.; Or, M.E.; et al. Use of Hair as Matrix for Trace Elements Biomonitoring in Cattle and Roe Deer Sharing Pastures in Northern Italy. *Animals* **2024**, *14*, 2209. [CrossRef]
19. Sander, J.; Terhardt, M.; Janzen, N.; Renaud, B.; Kruse, C.J.; Francois, A.C.; Wouters, C.P.; Boemer, F.; Votion, D.M. Tissue Specific Distribution and Activation of Sapindaceae Toxins in Horses Suffering from Atypical Myopathy. *Animals* **2023**, *13*, 2410. [CrossRef] [PubMed]
20. Renaud, B.; Kruse, C.J.; François, A.C.; Cesarini, C.; van Loon, G.; Palmers, K.; Boemer, F.; Luis, G.; Gustin, P.; Votion, D.M. Large-scale study of blood markers in equine atypical myopathy reveals subclinical poisoning and advances in diagnostic and prognostic criteria. *Environ. Toxicol. Pharmacol.* **2024**, *110*, 104515. [CrossRef]

21. Mee, J.F.; Boyle, L.A. Assessing whether dairy cow welfare is "better" in pasture-based than in confinement-based management systems. *N. Z. Vet. J.* **2020**, *68*, 168–177. [CrossRef] [PubMed]
22. Arnott, G.; Ferris, C.P.; O'Connell, N.E. Review: Welfare of dairy cows in continuously housed and pasture-based production systems. *Animal* **2017**, *11*, 261–273. [CrossRef] [PubMed]
23. Hitchens, P.L.; Hultgren, J.; Frössling, J.; Emanuelson, U.; Keeling, L.J. An epidemiological analysis of equine welfare data from regulatory inspections by the official competent authorities. *Animal* **2017**, *11*, 1237–1248. [CrossRef] [PubMed]

animals MDPI

Article

Unravelling Faecal Microbiota Variations in Equine Atypical Myopathy: Correlation with Blood Markers and Contribution of Microbiome

Anne-Christine François [1,*,†], Carla Cesarini [2,†], Bernard Taminiau [3], Benoît Renaud [1], Caroline-Julia Kruse [4], François Boemer [5], Gunther van Loon [6], Katrien Palmers [7], Georges Daube [3], Clovis P. Wouters [1], Laureline Lecoq [2], Pascal Gustin [1] and Dominique-Marie Votion [1]

1 Department of Functional Sciences, Faculty of Veterinary Medicine, Pharmacology and Toxicology, Fundamental and Applied Research for Animals & Health (FARAH), University of Liège, 4000 Liège, Belgium; benoit.renaud@uliege.be (B.R.); p.gustin@uliege.be (P.G.); dominique.votion@uliege.be (D.-M.V.)
2 Equine Clinical Department, Faculty of Veterinary Medicine, Fundamental and Applied Research for Animals & Health (FARAH), University of Liège, 4000 Liège, Belgium; ccesarini@uliege.be (C.C.); laureline.lecoq@uliege.be (L.L.)
3 Department of Food Sciences–Microbiology, Faculty of Veterinary Medicine, Fundamental and Applied Research for Animals & Health (FARAH), University of Liège, 4000 Liège, Belgium; bernard.taminiau@uliege.be (B.T.); georges.daube@uliege.be (G.D.)
4 Department of Functional Sciences, Faculty of Veterinary Medicine, Physiology and Sport Medicine, Fundamental and Applied Research for Animals & Health (FARAH), University of Liège, 4000 Liège, Belgium; caroline.kruse@uliege.be
5 Biochemical Genetics Laboratory, CHU, University of Liège, 4000 Liège, Belgium; f.boemer@chuliege.be
6 Department of Internal Medicine, Reproduction and Population Medicine, Faculty of Veterinary Medicine, Ghent University, 9820 Merelbeke, Belgium; gunther.vanloon@ugent.be
7 De Morette Equine Clinic, 1730 Asse, Belgium; katrien.palmers@demorette.be
* Correspondence: acfrancois@uliege.be
† These authors contributed equally to this work.

Simple Summary: Equine atypical myopathy is a severe intoxication caused by protoxins synthesised by certain maple trees, notably sycamore maple (*Acer pseudoplatanus*). These protoxins are activated into harmful toxins that disrupt lipid metabolism by inhibiting specific steps of fatty acid β-oxidation, leading to the accumulation of acylcarnitines in the blood. This activation process is catalysed mainly by specific mitochondrial enzymes, which are also present in some bacteria. Horses grazing in close proximity to affected animals have shown differences in their faecal microbiota composition, suggesting that the role of gut microbiota in atypical myopathy could be more substantial than previously understood. Recently, blood analyses have demonstrated the existence of subclinical cases among these cograzers. The present study compares the faecal bacteria of horses affected by atypical myopathy, their cograzers, and a group of toxin-free horses serving as a control group. Results show significant differences in faecal bacterial diversity and composition between groups, particularly for certain bacterial genera. Additionally, blood levels of specific compounds appear to be associated with these bacterial changes. The theoretical presence of the enzymes involved in the protoxin activation process was also studied. These results highlight the importance of comprehensively studying intestinal microbiota to better understand its role in this generally fatal poisoning.

Abstract: Hypoglycin A and methylenecyclopropylglycine are protoxins responsible for atypical myopathy in equids. These protoxins are converted into toxins that inhibit fatty acid β-oxidation, leading to blood accumulation of acylcarnitines and toxin conjugates, such as methylenecyclopropylacetyl-carnitine. The enzymes involved in this activation are also present in some prokaryotic cells, raising questions about the potential role of intestinal microbiota in the development of intoxication. Differences have been noted between the faecal microbiota of cograzers and atypical myopathy-affected horses. However, recent blood acylcarnitines profiling revealed subclinical cases among cograzers, challenging their status as a control group. This study investigates the faecal microbiota of horses clinically affected by atypical myopathy, their cograzers, and a control group of toxin-free horses while analysing correlations between microbiota composition and blood parameters. Faecal samples were analysed using 16S amplicon sequencing, revealing significant differences in α-diversity, evenness, and β-diversity. Notable differences were found between several genera, especially *Clostridia_ge*, *Bacteria_ge*, *Firmicutes_ge*, *Fibrobacter*, and *NK4A214_group*. Blood levels of methylenecyclopropylacetyl-carnitine and C14:1 correlated with variations in faecal microbial composition. The theoretical presence of enzymes in bacterial populations was also investigated. These results underscore the critical need to investigate the potential role of intestinal microbiota in this poisoning and may provide insights for developing prevention and treatment strategies.

Keywords: equine atypical myopathy; microbiota; gut microbiota; faecal microbiota; microbiome; faecal microbiome; horses; equine; hypoglycin A; methylenecyclopropylacetyl-carnitine; MCPA-CoA; acylcarnitines; 16S rRNA gene sequencing; next generation sequencing; NGS; blood metabolites; toxin; poisoning

1. Introduction

Equine atypical myopathy (AM) is a seasonal and highly fatal [1] intoxication caused by the ingestion and metabolism of two protoxins contained in some *Acer* spp., including *Acer pseudoplatanus* and *Acer negundo* [2,3]: methylenecyclopropylalanine, also known as hypoglycin A (HGA), [2] and methylenecyclopropylglycine (MCPrG) [4].

Both HGA and MCPrG are non-proteinogenic amino acids [5] that are not toxic *per se*. Their metabolism into toxic metabolites requires the action of enzymes, mainly located in the mitochondria, involved in branched-chain amino acids (BCAAs) catabolic pathway [6]. The first step is a transamination, catalysed by the branched-chain amino acid aminotransferase (BCAT) [6,7]. The second and irreversible step, catalysed by the branched-chain α-keto acid dehydrogenase complex (BCKDHc), is an oxidative decarboxylation with coenzyme A (CoA) added to oxidised products [6,8]. This results in the formation of methylenecyclopropylacetyl-CoA (MCPA-CoA) for HGA and methylenecyclopropylformyl-CoA (MCPF-CoA) for MCPrG [6]. These toxic metabolites impair the β-oxidation of fatty acids, leading to increased levels of acylcarnitine in tissues [9], urine [2,10,11] and blood [5,12–16] The diagnosis of AM relies on a combination of clinical signs, acylcarnitine profiling, and the detection of toxic metabolites conjugated with carnitine or glycine, such as MCPA-carnitine [3,16–19].

Both prokaryotic and eukaryotic cells can metabolise amino acids [20]. Indeed, some bacteria have the above-mentioned enzymes: for example, *Lactococcus lactis* and *Escherichia coli* express BCAT [21,22], and *Bacillus subtilis*, *Staphylococcus aureus* and *Pseudomonas putida* express BCKDHc [23–31]. This raises the hypothesis that the intestinal microbiota may play a role in the metabolism of HGA and/or MCPrG, potentially affecting horses that have ingested these protoxins. Analysis of the faecal microbiota is of twofold interest in exploring this hypothesis: on top of being a non-invasive technique, the faecal microbiota is reflective of the colonic microbial population in equids [32–34].

Wimmer-Scherr et al. (2021) conducted a study exploring the differences in faecal microbiota between horses with AM and a control group consisting of their cograzers (CoG)

(i.e., horses grazing in the same pasture as a poisoned horse but having a normal physical examination). They described that (1) the relative abundance of families *Ruminococcaceae*, *Christensenellaceae* and *Akkermansiaceae* was higher in horses suffering from AM compared to their CoG, and (2) the relative abundance of families *Lachnospiraceae*, *Bacteroidales* and *Clostridiales* was lower in horses with AM compared to their CoG, especially in non-surviving AM animals [35].

Clinically healthy CoG of HGA-poisoned horses are exposed to similar toxic pressure as horses affected by AM, yet they do not exhibit any clinical signs of poisoning [36–38]. Due to their shared environment and, therefore, exposure to the same protoxins, their blood often contains detectable levels of HGA and, in some cases, MCPA-carnitine [10,17,37,39] at levels that sometimes overlap with those observed in clinically affected AM horses [16,38]. Despite the absence of overt clinical signs, acylcarnitines profiling—which is the diagnostic and prognostic gold standard of AM [15,16]—revealed that many CoG has increased levels of acylcarnitines compared to control horses (i.e., horses free of toxin) [16]. This observation indicates the existence of subclinical cases among CoG, challenging their status as a healthy control group. Furthermore, another study on *Acer pseudoplatanus* poisoning in herbivorous species highlighted the existence of subclinical poisoning in species other than equids and hypothesised a potential role of gut morphology and intestinal microbiota in the risk of HGA intoxication [40]. Therefore, to study the potential role of microbiota in AM, it is necessary to include a control group composed of clinically healthy toxin-free horses and compare this group to both AM horses and their CoG.

As previously mentioned, some bacteria possess the enzymatic machinery necessary to metabolise HGA into MCPA. This raises the hypothesis that there could be a correlation between the host's microbiota and blood parameters related to intoxication (such as HGA, MCPA-carnitine and the acylcarnitines profile). This hypothesis was not addressed in the study of Wimmer-Scherr et al. (2021), highlighting a second point for improvement in understanding the possible role of intestinal microbiota in AM [35].

Therefore, the aims of this study were (1) to compare the faecal microbiota of AM horses (both survivors and non-survivors) and their CoG with that of control horses (i.e., grazing horses without protoxins or toxic metabolites in their blood) and (2) to analyse the correlation between faecal microbiota and blood parameters associated with AM intoxication in horses (i.e., HGA, MCPA-carnitine and the acylcarnitines profile).

2. Materials and Methods

All procedures in this study adhered to both national and international guidelines on animal welfare. The Animal Ethics Committee of the University of Liege was consulted, and it was confirmed that the sampling process was part of routine veterinary practice for diagnosing or preventing AM. As a result, formal ethical approval was not required. Informed consent was obtained from horse owners prior to their inclusion in the study.

2.1. Horses: Study Design, Inclusion Criteria and Group Definition

Four separate groups were defined for the purpose of this study:

- Control horses (CONTROL): clinically healthy horses free of HGA and MCPA-carnitine in the blood (i.e., HGA and MCPA-carnitine levels below the limit of detection), living in a place where AM cases had previously been observed and spending at least 6 h a day at pasture;
- Cograzers (CoG): clinically healthy horses grazing in a pasture where a case of AM had been diagnosed in the previous 24 h, which had a normal physical exam and a normal dynamic examination at walk (no signs of AM or other obvious disease) at the time of sampling;

- AM survivors (AM–S): horses diagnosed with AM that were discharged from the clinic once free of clinical signs after a variable hospitalisation period;
- AM non-survivors (AM–NS): horses diagnosed with AM that died from the intoxication during hospitalisation or had to be euthanised due to significant clinical deterioration or continuous or unmanageable pain leading to a poor prognosis [41].

These last two groups (i.e., AM-S and AM-NS) were merged into a single group, referred to as 'diseased horses', for certain statistical analyses to facilitate comparisons with previously published data [16].

The groups AM-NS, AM-S and CoG include published data from a prospective clinical study conducted from autumn 2016 until spring 2019 [35]. The horses from the CONTROL group were prospectively sampled in autumn 2020.

The diagnosis of AM was based on (1) the algorithm proposed by van Galen et al. (2012) [1] (i.e., a compatible history and clinical signs highly suggestive of AM during spring/autumn) and on (2) the presence of HGA and MCPA-carnitine in serum, a modified acylcarnitines profile compatible with the diagnosis of AM, and elevated serum activities of creatine kinase (CK) when available. For inclusion in the present study, only horses with available data on blood levels of HGA, MCPA-carnitine, and acylcarnitines were considered [14–16].

2.2. *Comparison of Faecal Microbiota Between Groups*

A fresh faecal sample was collected at the time of clinical admission for horses suspected of AM. Faecal samples of CoG of these diseased horses were taken within 24 h of the first horse in the shared pasture displaying clinical signs of AM.

The centre of a faecal ball was sampled after direct collection from the rectum or from a pile of recently passed faeces (<30 min) as described by Stewart et al. (2018) [42] and was directly placed in a conservation medium (Stool DNA stabiliser, PSP® Spin Stool DNA Plus Kit 00310, Invitek, Berlin, Germany) and stored at −20 °C until total bacterial DNA extraction.

2.2.1. Bacterial DNA Extraction and High-Throughput Sequencing

The PSP Spin Stool DNA Plus Kit 00310 (Invitek, Berlin, Germany) was used to extract total bacterial DNA from stool samples as recommended by the manufacturer. The following primers (with Illumina overhand adapters), forward (5′-GAGAGTTTGATYMTGGCTCAG-3′), and reverse (5′-ACCOGCOGGCTGCTGGCAC-3′) were used to perform PCR amplification of the 16S rDNA V1–V3 hypervariable region and library preparation. Each PCR product was purified with the Agencourt AMPure XP bead kit (Beckman Coulter, Pasadena, CA, USA) and subjected to a second PCR round for indexing using Nextera XT index primers 1 and 2. After purification, PCR products were quantified using the Quant-IT PicoGreen (ThermoFisher Scientific; Waltham, MA, USA) and diluted to 10 ng/µL. A final quantification of each library was performed using the KAPA SYBR® FAST qPCR Kit (KapaBiosystems; Wilmington, MA, USA) before normalisation, pooling and sequencing on a MiSeq sequencer using V3 reagents (Illumina; San Diego, CA, USA). Commercial Mock community positive controls using DNA from 10 defined bacterial species (ATCC MSA-1000, ATCC, Manassas, VI, USA) and negative controls (from extraction and PCR steps) were included in the sequencing run [43].

Raw amplicon sequencing libraries were submitted to the NCBI database under bioproject number PRJNA1170059.

2.2.2. Sequence Analysis and 16S rDNA Profiling

Sequence read processing was performed as previously described [43] using the MOTHUR software package v1.48 [44] and the VSEARCH algorithm for chimera detection [45]. For operational taxonomic unit (OTU) generation, a clustering distance of 0.03 was used. 16S reference alignment and taxonomical assignment, from phylum to genus, were performed with MOTHUR and were based upon the SILVA database (v1.38.1) of full-length 16S rDNA sequences [46].

2.2.3. Data Analysis

Subsampled datasets with 10,000 cleaned reads per sample were obtained and used to evaluate α-diversity and β-diversity using a vegan package (v 2.6-6.1) [47].

The analysis of α-diversity (i.e., measuring diversity within the community) and β-diversity (i.e., measuring diversity between communities or within the same community at different time points by considering sequence abundances or by considering only the presence–absence of sequences) are used to assess the ecology of a microbial community [48,49]. Indicators of α-diversity include the Chao richness index, reciprocal Simpson microbial diversity, and Simpson-derived evenness. Richness quantifies the number of species present within a community, while evenness describes how uniformly individuals are distributed among the species, highlighting the presence or dominance of certain species [50].

Differences in α-diversity between groups (AM-NS, AM-S, CoG, CONTROL) were evaluated with an ANOVA test followed by paired post hoc tests corrected with a two-stage linear step-up procedure of Benjamini, Krieger and Yekutieli using PRISM 10 (GraphPad Software; San Diego, CA, USA). Differences were considered significant for a p and q-value of 0.05 or less.

The β-diversity was analysed using vegan and vegan3d packages (v 1.3-0) [51] in R. Sample β-diversity was visualised with a Bray–Curtis dissimilarity matrix-based non-parametric dimensional scaling (NMDS) model. Differences between groups for the sample clustering and β-dispersion were assessed with analysis of variance (Adonis2), using Bray–Curtis dissimilarity matrices (Adonis2) tests, and post hoc paired tests (pairwise adonis package v 0.4.1) [52] with p threshold of 0.05, using R studio.

The differential abundance analysis of the genus populations of the whole system was carried out using the Aldex function (unpaired test in ALDEX2 package v1.36.0) [53]. The Monte Carlo method used in Aldex involves generating random samples from the dataset and calculating the statistics of interest for each sample. By repeating this process, an empirical estimate of the null statistical distribution is obtained—that is, the distribution assuming no significant difference between the randomly generated samples. The p-values can then be calculated by comparing the observed statistics with this empirical null distribution.

A differential abundance analysis was performed with the Deseq2 package in R (v1.44.0) to highlight statistical differences in population abundance between pairs of groups.

Finally, the theoretical presence of BCAT (EC 2.6.1.42) and BCKDHc (E_1: EC 1.2.4.4–E_2: EC 2.3.1.168.–E_3: EC 1.8.1.4. for the three different subunits respectively) in bacterial profiles were investigated with Picrust2 tool [54]. This tool allows for theoretical metagenome function prediction based on 16S sequences. Orthologous sequences corresponding to these enzymes were identified using the Kyoto Encyclopaedia of Genes and Genomes (KEGG) database (ver. 2024-11-20, http://www.kegg.jp/kegg/, accessed on 23 November 2024). The presence of these orthologous sequences in the genus populations from our dataset was identified from known genome content present in the database. The statistical analyses and graphical representations of the number of pseudo-counts of each ortholog identified by groups of horses were performed using GraphPad Prism 10 (GraphPad Software; San Diego, CA, USA). The Kolmogorov-Smirnov test was used to assess the normality or lognormality of the data distribution, followed by a Kruskal-Wallis multiple comparisons test.

2.3. Correlation Between Faecal Microbiota and Blood Parameters

A blood sample was collected from each diseased horse upon admission to the clinic. Cograzers of these diseased horses were sampled in the field within 24 h after the first affected horse in the pasture showed clinical signs of AM. Blood samples were obtained via jugular venipuncture, aliquoted within one hour of collection, and stored at −80 °C until analysis. Both blood samples and faecal samples were collected simultaneously from each horse.

2.3.1. Quantification of Hypoglycin A, Methylenecyclopropylacetyl-Carnitine and Acylcarnitines

The quantification of HGA was performed on serum according to a previously described methodology using aTRAQ® kit (Sciex, Framingham, MA, USA) for amino acid analysis of physiological fluids [17]. Briefly, HGA was derivatised using an isotopic tag (mass m/z 121), while a second labelling reagent (mass m/z 113) allowed absolute quantification. The samples were derivatised and introduced into a TQ5500 tandem mass spectrometer (Sciex, Framingham, MA, USA) using a Prominence AR HPLC system (Shimadzu, Kyoto, Japan). The lower limit of quantification associated with this method is 0.090 µmol/L and a coefficient of variation below 8% [17].

An ultra-performance liquid chromatography combined with subsequent tandem mass spectrometry (UPLC-MS/MS) was used for MCPA-carnitine quantification with a limit of detection of approximately 0.001 nmol/L, as previously described [2].

Free carnitine and acylcarnitine profiling were quantified in serum by tandem mass spectrometry. Serum proteins were precipitated with a methanol solution with labelled internal standards. After evaporation with a nitrogen stream and derivatisation with butanolic-HCl, the samples were analysed with a TQ5500 mass spectrometer (Sciex, Framingham, MA, USA) [15,55].

2.3.2. Statistical Analysis of Group Parameters and Blood Markers Associated with Intoxication

The statistical analyses and graphical representations were performed using GraphPad Prism 10 (GraphPad Software; San Diego, CA, USA). The Kolmogorov-Smirnov test was employed to assess the assumption of normal or lognormal distribution: the data were normalised through Log10 transformation if needed.

A comparison of the means of the parameters "Age" and "Sex" and the blood concentration of HGA and MCPA-carnitine was conducted by comparing groups as previously described in the literature [16].

The acylcarnitines profiling was studied by comparing the averages between groups of horses and by focusing on the recently published cutoffs isovaleryl-/2-methylbutyrylcarnitine (i.e., C5 acylcarnitine) [13,15,16].

2.3.3. Statistical Analysis of the Correlation Between Faecal Microbiota and Blood Parameters

A Mantel test was performed to assess the correlation between faecal microbiota and blood parameters related to intoxication: HGA, MCPA-carnitine and a selection of acylcarnitines (i.e., C2, C4, C5, C10, C12:1, C14, C14:1, and C18:1) in serum, based on previous publications [15,16].

A multivariate analysis of the influence of selected acylcarnitines, HGA and MCPA-carnitine on microbiota distribution was performed using a distance-based redundancy analysis (dbRDA) with the vegan package in R software. This dbRDA was performed (1) to explore the interactions between the chemical variables (i.e., HGA, MCPA-carnitine and selected acylcarnitines) and the microbial composition of samples, (2) to identify the most important variables, (3) to visualise results, and (4) to statistically validate relationships between variables and sample composition. The significance of parameter influence on the dimensional model was assessed with ANOVA (vegan package). A p-value of 0.05 or less was considered statistically significant. The dbRDA model was illustrated with the ggord package in R (v1.1.8) [56].

3. Results

3.1. Horses

Changes in the classification of bacteria since 2021 justified a re-analysis of faecal samples from AM horses and CoG coming from Wimmer-Scherr and collaborators' study [35]: during this process, one animal was discarded due to lack of bacterial DNA in the faecal samples to allow further analysis.

A total of 36 horses were included: 13 horses in the AM-NS, 12 horses in the AM-S, 5 horses in the CoG, and 6 horses in the CONTROL group.

The final population included a mix of different breeds: Andalusians, Belgian Warmblood Horses, Belgian Draft Horses, Friesians, Haflingers, Hanoverians, Irish Cobs, Merens, Ponies, Quarter Horses, French saddlebreds, Trotters and Zangersheides. The age and sex distribution of each group are presented in Table 1.

Table 1. Demographic data of horses classified by group.

	CONTROL	CoG	AM-S	AM-NS	Diseased Horses
Number of horses	6	5	12	13	25
Age mean ± SD (Years)	13.3 ± 8.5	7.1 ± 7.5	6.5 ± 7.7	5.8 ± 4.7	6.1 ± 6.2
Age minimum (Years)	4.0	1.5	0.4	0.5	0.4
Age maximum (Years)	23	17	25	16	25
Age CI [LLCI-ULCI]	[6.5–20.1]	[0.5–13.7]	[2.1–10.9]	[3.3–8.3]	[3.7–8.5]
Ratio of					
entire male	33%	40%	33%	15%	24%
gelding	33%	20%	33%	23%	48%
female	33%	40%	33%	62%	28%

SD = Standard deviation, CI = Confidence interval, LLCI = Lower Limit of Confidence Interval, ULCI = Upper Limit of Confidence Interval. The groups represented are control horses (CONTROL), cograzers (CoG), survivors (AM-S) and non-survivors (AM-NS) of atypical myopathy, as well as all diseased horses (AM-S + AM-NS).

3.2. Comparison of Faecal Microbiota Between Groups

3.2.1. Composition of Faecal Microbiota

Starting with 6,099,718 raw reads, 3,649,756 reads were kept after read cleaning and chimera removal. We further proceeded with 10,000 reads per sample to the taxonomic identification, leading to a table of 184,346 OTUs.

Among the 21 phyla identified in the faeces, the 4 most abundant defined phyla were Firmicutes, Bacteroidota, Verrucomicrobiota and Fibrobacterota. A total of 206 families and 440 genera were identified. The dominant bacterial populations for each group and individual are detailed in Appendix A. Figures show results by group and for each individual in terms of phyla (Figure A1), family (Figure A2) and genera (Figure A3). The determination of α- and β-diversities of the faecal bacterial populations were assessed at the genus level.

3.2.2. α-Diversity Analysis

A test of normality was performed, and a normal distribution of data was indicated. Consequently, an ANOVA and a paired test were performed ($p = 0.05$). The outcomes of Benjamini, Krieger and Yekutieli with a false discovery rate (FDR) ($q = 0.05$) indicated significant disparities among the groups concerning bacterial α-diversity (reciprocal Simpson Index) and evenness (i.e., distribution of abundances of the groups—Simpson Evenness index), though not for richness (i.e., number of taxonomic groups—chao1 richness index) (Figure 1).

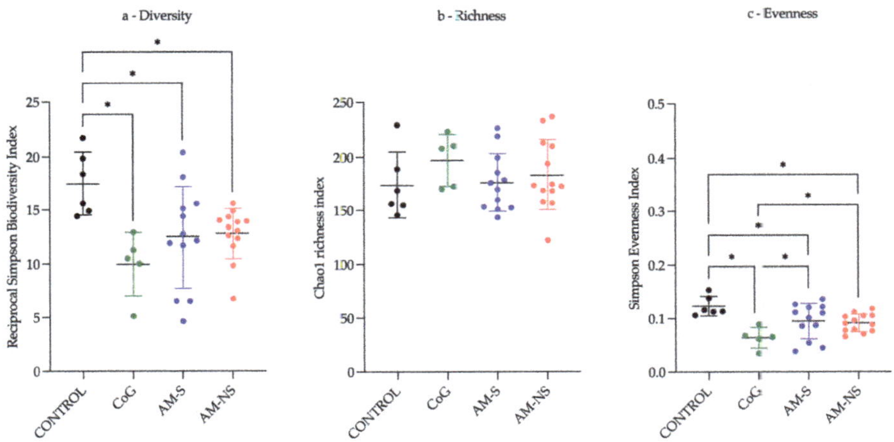

Figure 1. Representation of bacterial intrinsic diversity deduced from inverse Simpson index ((**a**)—Diversity), bacterial genus richness deduced from Chao1 index ((**b**)—Richness) and bacterial genus evenness deduced from Simpson index ((**c**)—Evenness). Data are scatter dot plots at the genus level for individual horses in the defined groups (control horses (CONTROL), cograzers (CoG), survivors (AM-S) and non-survivors (AM-NS) among atypical myopathy cases), with the mean and the standard deviation. The two-stage linear step-up procedure of Benjamini, Krieger and Yekutieli reveals significant differences with a q-value of 0.05 or less: * < 0.05.

Pairwise comparisons showed that α-diversity was significantly higher in CONTROL horses than in CoG ($q < 0.01$), in AM-S ($q < 0.01$), and in AM-NS ($q < 0.05$). Similarly, genus evenness was found to be (1) significantly higher in the CONTROL group vs. CoG ($q < 0.001$), AM-S ($q < 0.05$), and AM-NS ($q < 0.05$) and (2) significantly lower in the CoG group vs. AM-S ($q < 0.05$) and AM-NS ($q < 0.05$) (Table 2).

Table 2. α-diversity: p-value for pairwise comparisons and q-value for the two-stage linear step-up procedure of Benjamini, Krieger and Yekutieli.

DIVERSITY			***p*-Value**	***q*-Value**
CONTROL	**vs.**	**CoG**	**	**
CONTROL	**vs.**	**AM-S**	**	**
CONTROL	**vs.**	**AM-NS**	*	*
CoG	vs.	AM-S	nsig	nsig
CoG	vs.	AM-NS	nsig	nsig
AM-S	vs.	AM-NS	nsig	nsig

Table 2. *Cont.*

EVENNESS			*p*-Value	*q*-Value
CONTROL	**vs.**	**CoG**	***	***
CONTROL	**vs.**	**AM-S**	*	*
CONTROL	**vs.**	**AM-NS**	*	*
CoG	**vs.**	**AM-S**	*	*
CoG	**vs.**	**AM-NS**	*	*
AM-S	vs.	AM-NS	nsig	nsig

Significantly different with a *p*- or *q*-value of 0.05 or less: * < 0.05; ** < 0.01; *** < 0.001, nsig: Not significant. The defined groups are control horses (CONTROL), cograzers (CoG), survivors (AM-S) and non-survivors (AM-NS) of atypical myopathy.

3.2.3. β-Diversity Analysis

β-diversity of the faecal microbial profile was visualised using a Bray-Curtix matrix-based NMDS model (k = 3, stress = 0.082) (Figure 2). Sample clustering showed that bacterial profiles were not homogenous between groups ($p = 0.001$). Results from paired tests ($p = 0.05$) showed that the microbial profile of the CONTROL group was different from the other groups, and CoG was different from AM-NS, as shown in Table 3.

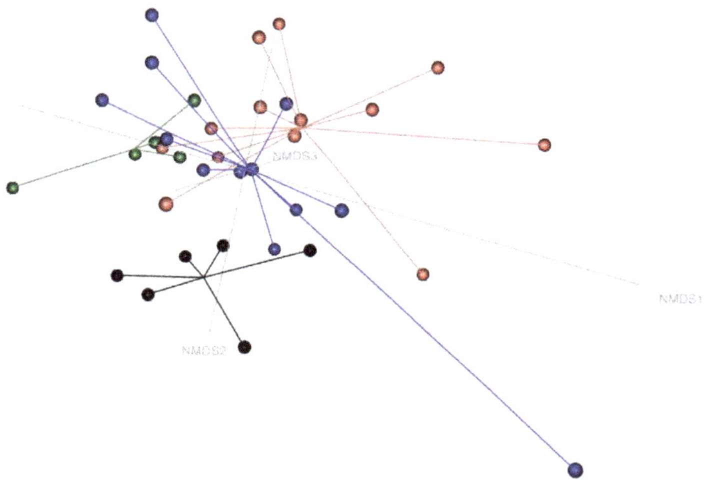

Figure 2. Nonmetric multidimensional scaling model (NMDS) plots in three dimensions of the horse's faecal microbiota. Black symbols represent control horses (CONTROL), green symbols represent cograzers (CoG), blue symbols represent survivors of atypical myopathy (AM-S), and red symbols represent non-survivors (AM-NS). The model stress is 0.082.

Table 3. β-diversity: *p*-value adjusted for pairwise comparisons.

	Pairs		*p*-Value Adjusted
CONTROL	**vs.**	**CoG**	*
CONTROL	**vs.**	**AM-S**	*
CONTROL	**vs.**	**AM-NS**	**
CoG	vs.	AM-S	nsig
CoG	**vs.**	**AM-NS**	*
AM-S	vs.	AM-NS	nsig

Significantly different with a *p*-value of 0.05 or less: * < 0.05; ** < 0.01; nsig: Not significant. The defined groups are control horses (CONTROL), cograzers (CoG), survivors (AM-S) and non-survivors (AM-NS) of atypical myopathy.

3.2.4. Differences in Global Faecal Microbiota Composition

The Aldex function revealed six significantly different genera in the samples studied (without the notion of group), compared with the null statistical distribution empirically determined by this aldex function: *Clostridia_ge*, *Bacteria_ge*, *Firmicutes_ge*, *Phascolarctobacterium*, *Fibrobacter* and *NK4A214_group* (Table 4).

Table 4. Aldex function: adjusted *p*-value of selected genera.

Genera	*p*-Value Adjusted
Clostridia_ge	**
Bacteria_ge	**
Firmicutes_ge	**
Phascolarctobacterium	*
Fibrobacter	*
NK4A214_group	*

Significantly different with a *p*-value of 0.05 or less: * < 0.05; ** < 0.01.

3.2.5. Differences in Faecal Microbiota Composition Between Groups

At a genus level, the statistically significant differences in population abundance between the CONTROL group and the three other groups (CoG, AM-S and AM-NS, respectively) are represented in Table 5. From a total of 34 genera, 4 genera were statistically different between CONTROL and CoG, 12 genera were statistically different between CONTROL and AM-S, and 32 genera were statistically different between CONTROL and AM-NS. These genera are graphically represented in Figure 3.

The 4 genera statistically different between CONTROL and CoG were also different between CONTROL and AM-S and AM-NS. Interestingly, the relative abundance of these four genera was significantly lower in CONTROL vs. CoG, AM-S and AM-NS. These genera are *Firmicutes_ge*, *Clostridia_ge*, *Bacteria_ge* and *Oligosphaeraceae_ge*. Three of them (i.e., *Firmicutes_ge*, *Clostridia_ge* and *Bacteria_ge*) were also highlighted by the Aldex function. The genera *NK4A214_group* and *Fibrobacter*—also revealed by the Aldex function—presented a significantly lower and higher relative abundance between CONTROL and AM-NS, respectively.

Table 5. Adjusted *p*-value corresponding to comparisons between groups.

Adjusted *p*-Value	CONTROL vs. CoG	CONTROL vs. AM-S	CONTROL vs. AM-NS
Firmicutes_ge	****	****	****
Clostridia_ge	****	****	****
Bacteria_ge	**	****	***
Oligosphaeraceae_ge	*	***	**
Lachnospirales_ge		**	*
Bacilli_ge		**	**
Oscillospirales_ge		**	**
RF39_ge		*	
Catenibacillus		*	*
Treponema		*	*
Anaerovoracaceae_ge		*	
Anaeroplasma		*	**
Streptococcus			***
Bradymonadales_ge			***
Candidatus_Soleaferrea			***

Table 5. *Cont.*

Adjusted *p*-Value	CONTROL vs. CoG	CONTROL vs. AM-S	CONTROL vs. AM-NS
Oscillospiraceae_ge			**
Candidatus_Saccharimonas			**
Prevotellaceae_UCG.001			**
Akkermansia			**
Clostridiaceae_ge			**
NK4A214_group			**
Christensenellaceae_R.7_group			*
Prevotellaceae_UCG.003			*
Gastranaerophilales_ge			*
Campylobacter			*
Selenomonadaceae_ge			*
Prevotella			*
Bacteroidales_ge			*
Endomicrobium			*
COB_P4.1_termite_group_ge			*
Verrucomicrobiota_ge			*
Fibrobacter			*
Muribaculaceae_ge			*
Bacteroides			*

Significantly different with a *p*-value of 0.05 or less: * < 0.05; ** < 0.01; *** < 0.001, **** < 0.0001. The defined groups are control horses (CONTROL), cograzers (CoG), survivors (AM-S) and non-survivors (AM-NS) of atypical myopathy. The genera that were also highlighted by the Aldex function are underlined.

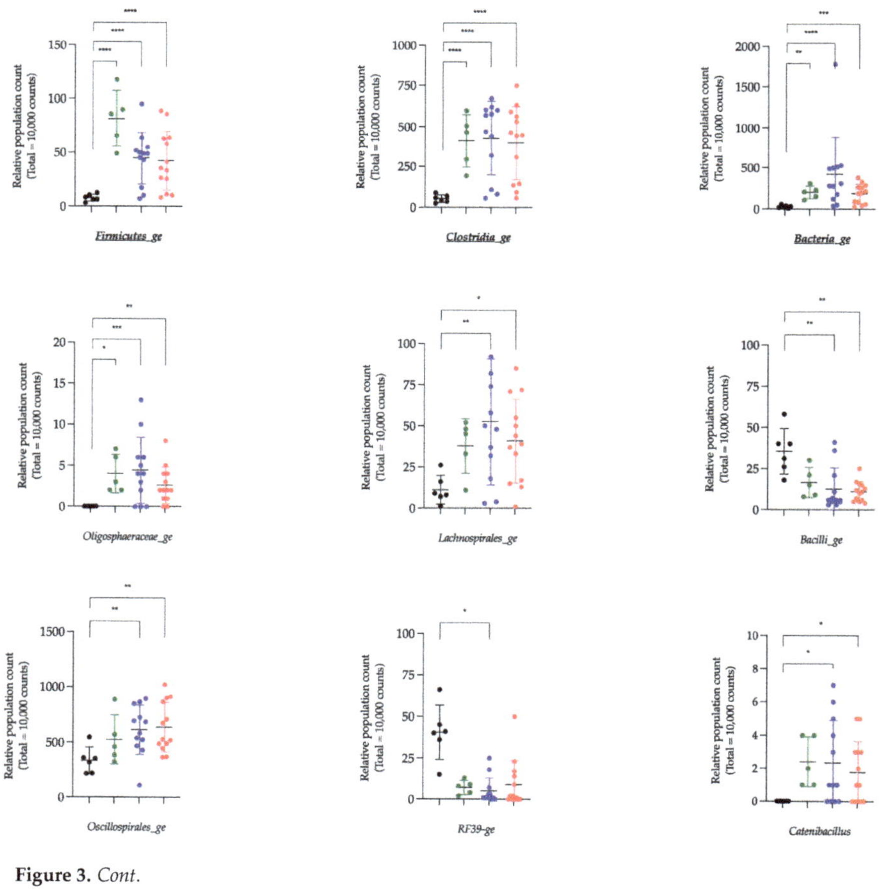

Figure 3. *Cont.*

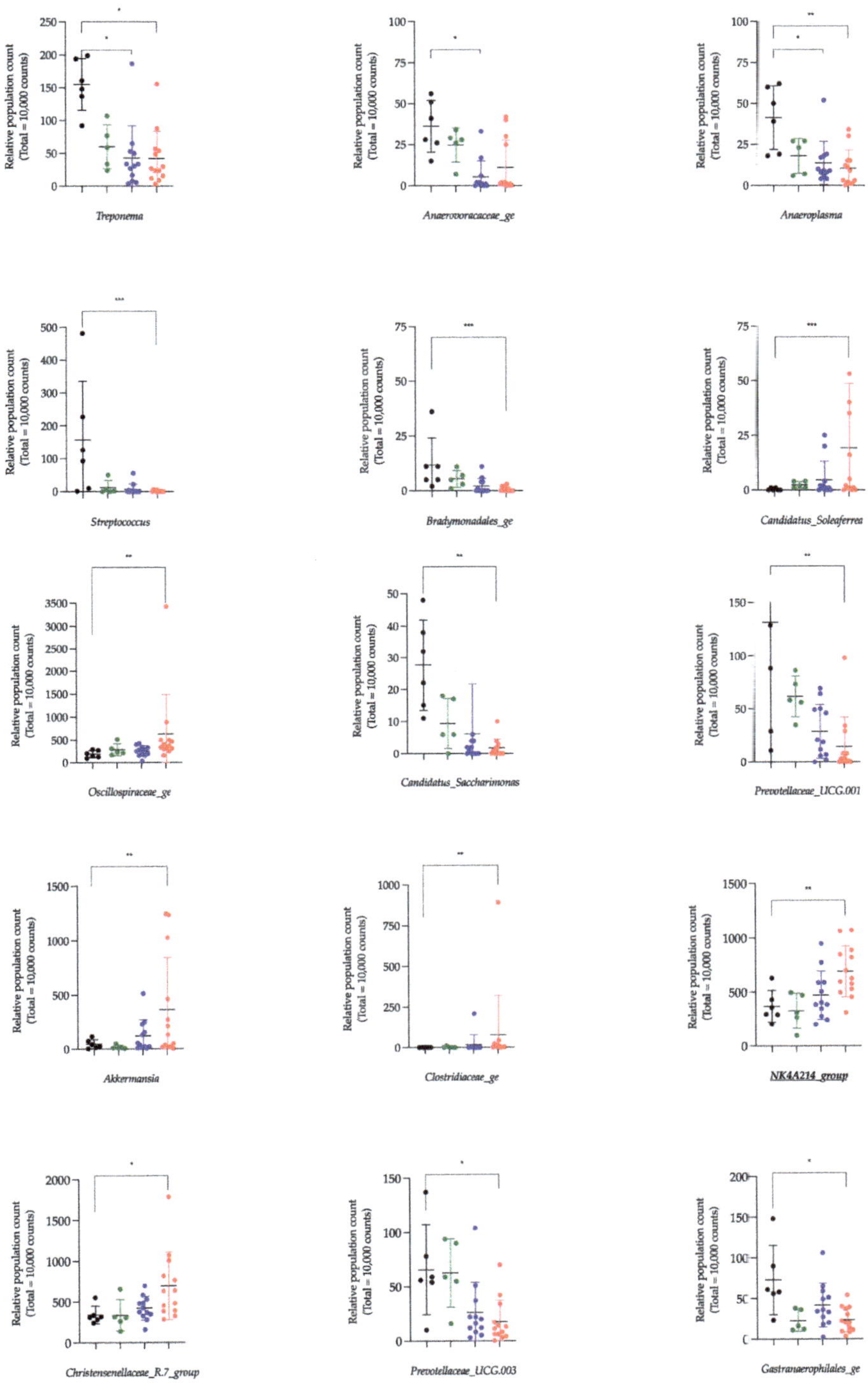

Figure 3. *Cont.*

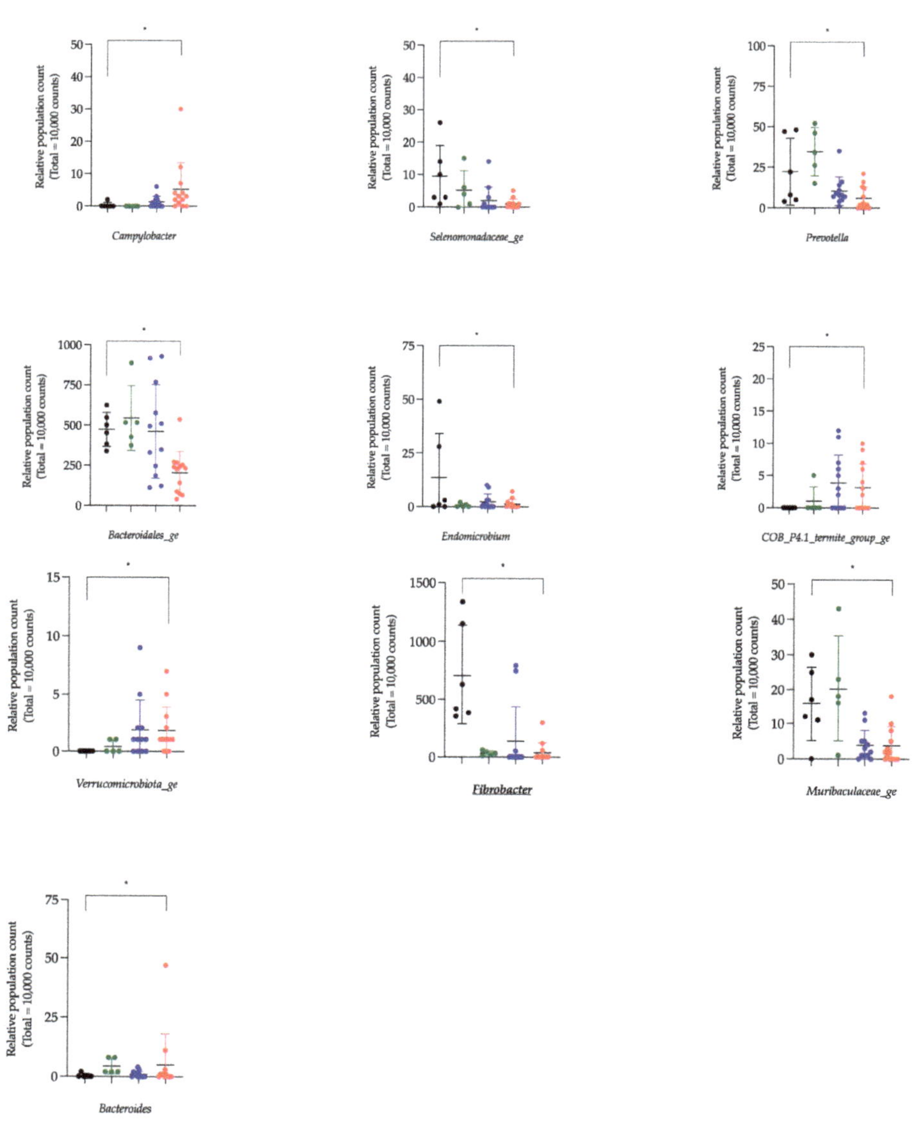

Figure 3. Graphical representation of relative population count (mean and standard deviation, total of 10,000 counts). Data are scatter dot plots at the genus level for individual horses in the defined groups, with black dots for control horses (CONTROL), green dots for cograzers (CoG), blue dots for survivors (AM-S) and red dots for non-survivors (AM-NS) of atypical myopathy. Genus names in bold and underlined were also highlighted by the Aldex function. Significantly different with a p-value of 0.05 or less: * < 0.05; ** < 0.01; *** < 0.001, **** < 0.0001.

3.2.6. Enzymes of the Kyoto Encyclopaedia of Genes and Genomes Orthologous Pseudo-Counts Analysis

The KEGG orthologous for BCAT (EC 2.6.1.42) is K00826. For the BCKDHc, the subunit E1 presents three KEGG orthologous: K00166 for 2-oxoisovalerate dehydrogenase E1 component subunit alpha, K00167 for 2-oxoisovalerate dehydrogenase E1 component subunit beta and, K11381 for 2-oxoisovalerate dehydrogenase E1 component. The KEGG orthologous for subunits E2 and E3 are K09699 and K00382, respectively.

Using the Picrust2 tool, the repartition of pseudo-counts of each KEGG Orthology number (KO) by groups of horses is represented in Figure 4. As observed and expected, the Kruskal-Wallis analysis of the different pseudo-counts of each K0 did not reveal any statistical differences between groups of horses (i.e., CONTROL, CoG, AM-S and AM-NS).

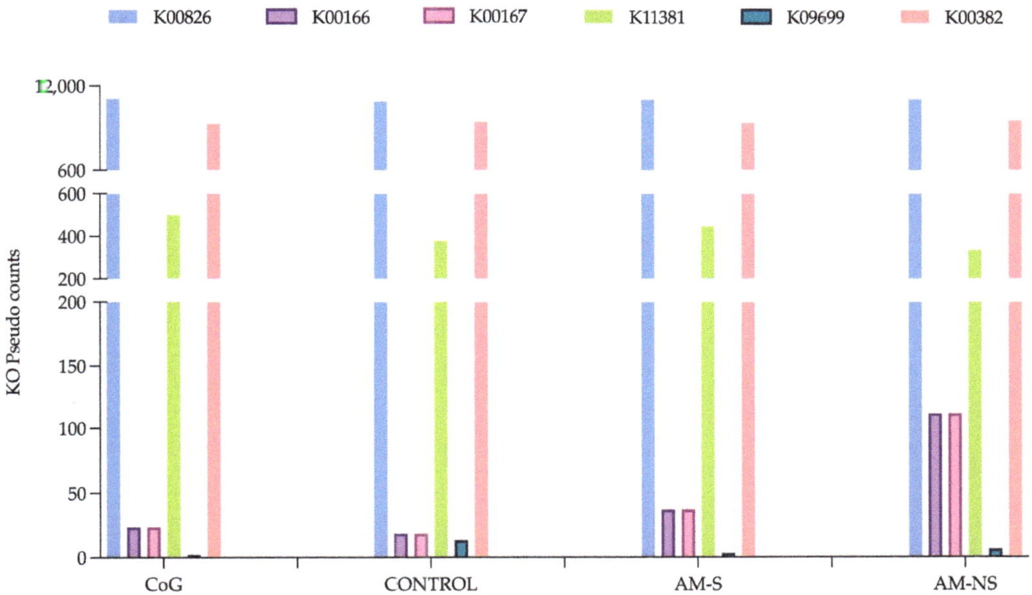

Figure 4. Graphical representation of each KO pseudo-counts in the defined groups: control horses (CONTROL), cograzers (CoG), survivors (AM-S) and non-survivors (AM-NS) of atypical myopathy.

The general analysis of the presence of KO in each OTU indicated that (1) K00826 (i.e., BCAT) was mainly represented in 99.18% of total OTUs, (2) K00382 (i.e., BCKDHc—E3) was represented in 70.13% of total OTUs and, (3) BCKDHc E1 and E2 subunits were the main limiting subunits. The latter were poorly represented within the entire 16S rDNA (0.35%, 0.35%, 3.28% and 0.08% for K00166, K00167, K11381 and K09699, respectively), with only four OTUs possessing all the enzymes, and these same four OTUs were the only ones to possess the entire E1 and E2 subunits. Moreover, the relative abundance of these four OTUs was low. The identification of the genus of these four OTUs was *Alphaproteobacteria_ge*, *Rhizobiaceae_ge*, *Sphingomonas*, and *Sphingomonadaceae_ge*.

Among the bacterial genera highlighted by Aldex and/or Deseq2 functions, the mostly represented are K00826 (i.e., BCAT) and K00382 (i.e., BCKDHc—E3), which agrees with the general analysis (Table 6). Moreover, many of the genera analysed only presented these two orthologs. *Phascolarctobacterium* did not present any pseudo-count for any KO number like some OTUs from some genera (i.e., *Anaerovoracaceae_ge*, *Bacilli_ge*, *Bacteria_ge*, *Bacteroidales_ge*, *Christensenellaceae_R.7_group*, *Clostridia_ge*, *Firmicutes_ge*, *Lachnospirales_ge*, *Oscillospirales_ge*). None of the genera identified possessed all the orthologs. Nevertheless, some OTUs from the genera *Bacilli_ge* and *Bacteria_ge* presented 5/6 orthologous sequences with the same missing K11381 (i.e., 2-oxoisovalerate dehydrogenase E1 component). Some OTUs from the genera *Akkermansia*, *Bacteria_ge* and *Verrucomicrobiota_ge* owned 4/6 orthologous sequences, with K11381 and K09699 missing (i.e., BCKDHc E2 subunit). Finally, one OTU from *Verrucomicrobiota_ge* and *Muribaculaceae_ge*, three OTUs from *Prevotella*, and several OTUs from *Bacteria_ge*, *Bacteroidales_ge*, *Bacteroides*, *Christensenellaceae_R.7_group*, *Clostridia_ge*, *COB_P4.1_termite_group_ge*, *Firmicutes_ge*, *NK4A214_group*, *Oscillospiraceae_ge*,

Oscillospirales_ge, *Prevotellaceae_UCG-001*, *Prevotellaceae_UCG-003* presented the same 3/6 orthologous sequences (i.e., K00826, K11381 and, K00382).

Table 6. The presence of orthologous sequences in bacterial genera is of interest.

Orthologs Number KEGG	K00826	K00166	K00167	K11381	K09699	K00382
Akkermansia	✔ ✔	✔	✔			✔ ✔
Anaeroplasma	✔					✔ ✔
Anaerovoracaceae_ge	✔					✔
Bacilli_ge	✔ ✔	✔	✔		✔	✔ ✔
Bacteria_ge	✔ ✔ ✔ ✔ ✔	✔ ✔	✔ ✔	✔	✔	✔ ✔ ✔ ✔
Bacteroidales_ge	✔ ✔ ✔			✔		✔ ✔
Bacteroides	✔ ✔			✔		✔ ✔
Bradymonadales_ge	✔					✔
Campylobacter	✔					
Candidatus_Saccharimonas	✔					✔
Candidatus_Soleaferrea	✔ ✔					✔
Catenibacillus	✔ ✔					✔
Christensenellaceae_R.7_group	✔ ✔ ✔			✔		✔ ✔
Clostridia_ge	✔ ✔ ✔			✔		✔ ✔ ✔
Clostridiaceae_ge	✔ ✔					✔
COB_P4.1_termite_group_ge	✔ ✔			✔		✔ ✔
Endomicrobium	✔					✔
Fibrobacter	✔ ✔					✔

Table 6. *Cont.*

Orthologs Number KEGG	K00826	K00166	K00167	K11381	K09699	K00382
Firmicutes_ge	✔			✔		✔
	✔					✔
	✔					
Gastranaerophilales_ge	✔					✔
Lachnospirales_ge	✔					✔
	✔					
Muribaculaceae_ge	✔			✔		✔
	✔					✔
NK4A214_group	✔			✔		✔
	✔					✔
	✔					
Oligosphaeraceae_ge	✔					✔
Oscillospiraceae_ge	✔			✔		✔
	✔					✔
	✔					
Oscillospirales_ge	✔			✔		✔
	✔					✔
	✔					
Phascolarctobacterium						
Prevotella	✔			✔		✔
	✔					✔
	✔					
Prevotellaceae_UCG-001	✔			✔		✔
	✔					✔
	✔					
Prevotellaceae_UCG-003	✔			✔		✔
	✔					✔
	✔					
RF39_ge	✔					✔
Selenomonadaceae_ge	✔					✔
	✔					
Streptococcus	✔					✔
Treponema	✔					✔
	✔					
Verrucomicrobiota_ge	✔	✔	✔			✔
	✔			✔		✔
	✔					✔

Unfilled rows within a bacterial genus represent one or more OTUs of this genus that do not possess any of the orthologues.

3.3. *Correlation Between Faecal Microbiota and Blood Parameters*

3.3.1. Group Parameters and Blood Markers Associated with Intoxication

The average blood concentrations of HGA, MCPA-carnitine and selected acylcarnitines, as well as their standard deviation for each group, are referenced in Tables A1 and A2.

Statistical comparison of the "Age" parameter between (1) CoG vs. diseased horses, (2) AM-S vs. AM-NS, (3) CONTROL vs. CoG and (4) CONTROL vs. diseased horses did not reveal any significant difference in mean age (Mann-Whitney test, $p \geq 0.05$) or in the age distribution (Kolmogorov-Smirnov Z test, $p \geq 0.05$).

Fisher's exact test used to study contingency about parameter "Sex" and parameter "Health Status" or "Final Outcome of the disease horses" did not reveal any significant difference ($p \geq 0.05$) for both comparisons.

The mean serum concentration of HGA was significantly lower in CoG vs. diseased horses (unpaired *t*-test, $p < 0.001$), but the difference was not significant for the comparison AM-S vs. AM-NS (unpaired *t*-test, $p \geq 0.05$) (Figure A4).

When comparing the mean serum concentration of MCPA-carnitine, diseased horses presented values significantly higher than CoG (unpaired *t*-test, $p < 0.0001$). Similarly, the mean serum concentration of MCPA-carnitine was significantly higher in AM-NS vs. AM-S (unpaired *t*-test, $p < 0.05$) (Figure A4).

The mean serum concentration of selected acylcarnitines in diseased horses was found to exceed the 99th percentile of the CONTROL group values. Additionally, the C5-carnitine concentration of each horse was analysed, according to what's been proposed by Renaud et al. (2024), to classify horses into the appropriate groups. A cutoff value of 3.04 µmol/L of C5 was described: above this cutoff, 92% of horses are diseased horses, and below this cutoff, 97% of horses are CoG. A second C5 cutoff is described with a value of 12.21 µmol/L: above this cutoff, 76% of diseased horses are likely to die, and below this cutoff, 81% of diseased horses are likely to survive [16]. In this study, only one AM-S horse (i.e., AM-S 04) presented a suspect profiling with a concentration of C5 below the cutoff of 3.04 µmol/L of clinically affected AM cases. Three AM-affected horses survived despite a poor prognosis based on C5 (Figure 5), and four diseased horses died while having a C5 level, suggestive of a positive outcome.

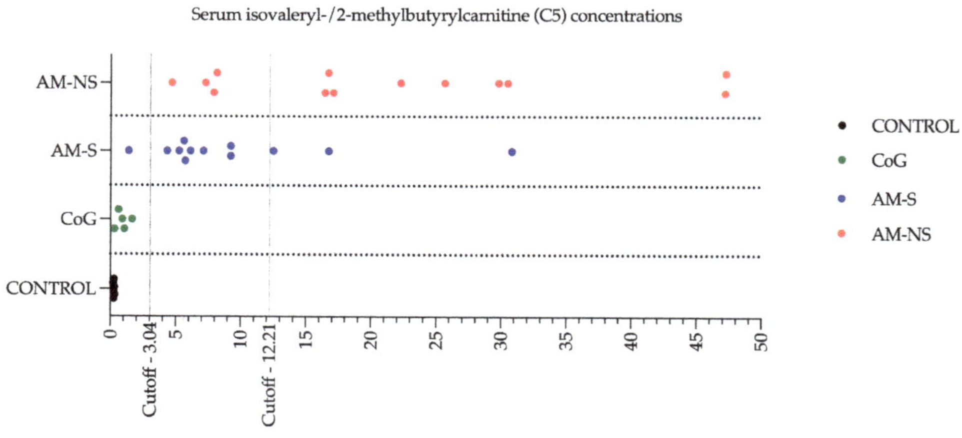

Figure 5. Serum isovaleryl-/2-methylbutyrylcarnitine (C5) concentrations. The groups are control horses (CONTROL), cograzers (CoG), survivors (AM-S) and non-survivors (AM-NS) of atypical myopathy.

3.3.2. Correlation Between Faecal Microbiota and Serum Concentration of Hypoglycin A, Methylenecyclopropylacetyl-Carnitine and Acylcarnitines

The Mantel test revealed a significant correlation ($r = 0.2351$, $p = 0.031$) between the matrix of microbiota and the matrix of blood parameters, suggesting that this correlation is unlikely due to chance.

A first dbRDA model was performed: the most important variables identified were MCPA-carnitine ($p = 0.004$), C2 ($p = 0.089$), C10 ($p = 0.099$) and C14:1 ($p = 0.079$).

A series of dbRDA models were further built iteratively by sequentially removing the significant chemical variables identified previously. This allowed us to select the most important variables to explain the variation in microbial composition between groups:

these variables were MCPA-carnitine ($p < 0.01$) and C14:1 ($p < 0.01$), resulting in a final dbRDA model with two constrained dimensions (Figure 6).

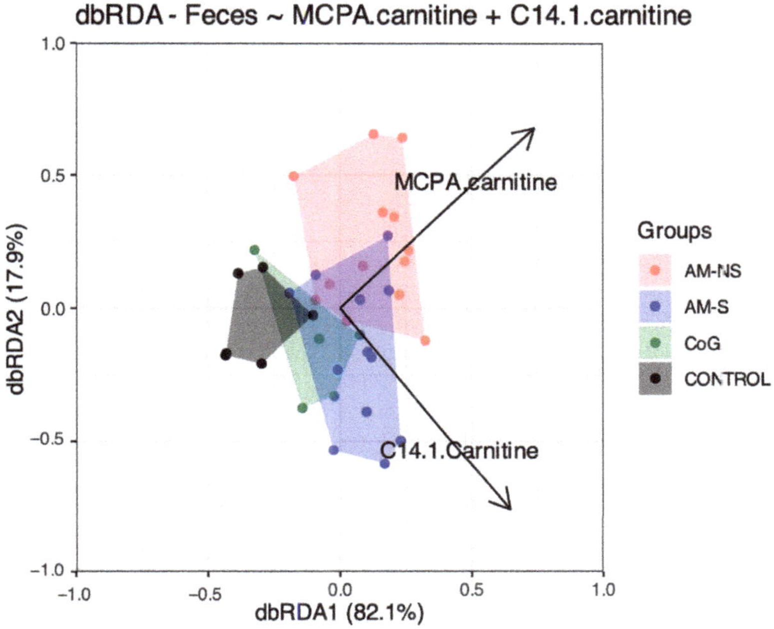

Figure 6. Illustration of the dbRDA model. The horizontal dbRDA1 axis explains 82.1%, and the vertical dbRDA2 axis explains 17.9% of the repartition of the different groups: control horses (CONTROL), cograzers (CoG), survivors (AM-S) and non-survivors (AM-NS) of atypical myopathy.

4. Discussion

The present study reveals that (1) faecal microbiota differs between CONTROL horses (i.e., toxin-free horses) and horses suffering from both subclinical (CoG) and clinical (AM-S and AM-NS) *Acer pseudoplatanus* intoxication and (2) the blood concentrations of MCPA-carnitine and C14:1 (i.e., tetradecenoylcarnitine, a long-chain acylcarnitine) significantly correlate with the variation in faecal microbial composition observed between the different groups of horses.

Microbiota statistical analyses were performed at the genus level to detect differences at the highest possible taxonomic resolution. This approach allows for a more accurate identification of the microbial populations influencing the system under investigation using the most up-to-date bacterial taxonomy. Unfortunately, recent updates in bacterial taxonomy since the publication of Wimmer-Scherr et al. (2021) have complicated a direct comparison of the results between the two studies [35]. Nevertheless, the findings of this new study will facilitate future comparisons with other research.

Regarding α-diversity, the CONTROL group presents the highest α-diversity and exhibits a more uniform distribution of populations compared to the three other groups, with an evenness index closer to one. The CoG group presents a lower α-diversity and an evenness index close to zero, indicating the emergence of certain populations more abundant than others. "Dysbiosis" is defined as the loss of central mutualistic relationship among microbiota members, metabolic products, and the host immune system [57]. Consequently, it is not possible to characterise the observed changes as dysbiosis only based on α-diversity indicators.

The Aldex function identified six genera having a significant impact in the global system (i.e., *Clostridia_ge, Bacteria_ge, Firmicutes_ge, Phascolarctobacterium, Fibrobacter* and *NK4A214_group*) compared to the null statistical distribution empirically determined by this function. The paired tests (Deseq2) compared bacterial genera between the CONTROL group and the three other groups and revealed 34 significantly different genera. Among these bacterial genera, five were also highlighted by the Aldex function. Indeed, *Clostridia_ge, Bacteria_ge,* and *Firmicutes_ge* presented a significantly lower relative abundance in the CONTROL group compared to CoG, AM-S and AM-NS groups, and *NK4A214_group* and *Fibrobacter* in the CONTROL group exhibit a significantly lower and higher relative abundance, respectively, compared to AM-NS horses.

The genera *Clostridia_ge, Bacteria_ge,* and *Firmicutes_ge* are composed of bacteria with features belonging to the *Clostridia* class, Bacteria kingdom or Firmicutes phyla, respectively, without the possibility of further characterisation. The genus *Fibrobacter* is a major and highly specialised cellulolytic bacterial genus and is detected in the intestinal tract of several herbivorous animals, including horses [58,59]. Cellulose is not digested and not absorbed in the mammal's gut, so cellulolytic bacteria play a vital role by providing energy to the host via the metabolism of cellulose into short-chain fatty acids (SCFAs). The SCFAs have beneficial effects on colonocytes, intestinal membrane integrity and local intestinal immunity [60,61]. The toxic metabolites of protoxins (i.e., MCPA-CoA and MCPF-CoA) inhibit the β-oxidation of fatty acids with subsequent decreased mitochondrial respiration and uncoupled phosphorylation [5,12,13]. Therefore, lipids can no longer be used as an energy substrate, but the glycolytic pathway is preserved [14,62,63]. Consequently, it is generally recommended to give a complete mix of grains or glucose in another form to horses affected by AM [64]. Although the appetite of diseased horses is preserved, the clinical picture (weakness, recumbency, depression, stiffness, etc. [1,36,65]) can make an inadequate ingestion of food. It is hypothesised that the inhibition of β-oxidation by toxic metabolites alters the overall energy metabolism of the host and the intestinal environment, potentially leading to conditions that are less favourable for the growth and activity of cellulolytic bacteria, thus explaining the observed decrease in the relative abundance of this genus from CONTROL horses to intoxicated horses (from CoG to AM-NS).

The genus *NK4A214_group* is associated with the degradation of structural carbohydrates, such as cellulose and hemicellulose, other less complex polysaccharide substrates and the production of butyrate (i.e., SCFAs) [66,67]. The relative increase in this genus from CONTROL horses to intoxicated horses, as previously reported [35], could be linked to the preservation of the glycolytic pathway and its ability to deal with less complex polysaccharides, as mentioned above. This preservation may provide a more favourable environment for the *NK4A214_group* by maintaining substrate availability for carbohydrate fermentation despite the metabolic disruptions caused by toxic exposure.

Other clearly identified genera, such as *Akkermansia, Bacteroides, Prevotella,* and *Treponema,* might be considered interesting in the context of AM. The genus *Akkermansia* attracts the attention of scientists for its anti-inflammatory and anti-obesity effects in mice and men by reducing insulin resistance, glucose intolerance, and gut permeability. One of the distinguishing features of *Akkermansia* is its ability to (1) promote the renewal and thickening of the mucin layer, which reduces intestinal permeability, and (2) degrade intestinal mucin glycoproteins and to use them as a source of carbon and nitrogen when dietary fibres are defective; this process leads to an increase of the intestinal permeability and to the production of SCFAs [68–72]. This ability to use mucin as a source of carbon and nitrogen may give *Akkermansia* an advantage when facing disruptions in host energy metabolism in the context of AM, despite the potential repercussions on intestinal microbiota. On the other hand, the resulting increase in intestinal permeability represents a disadvantage, as this could increase the absorption of HGA and other toxins present in the digestive tract and into the bloodstream. Lastly, the resulting production of SCFAs—which account for 65% of the horse's energy production [73,74]—cannot be used as an energy source by the horse. Indeed, in the case of AM, the muscle cannot use them effectively as energy substrates following the inhibition of acyl-CoA dehydrogenases [5,12,75–77].

The genus *Bacteroides*, which increases in AM-NS compared to CONTROL horses, is known for its ability to metabolise polysaccharides and oligosaccharides such as starch [78,79], which could make sense in the framework of the modification of the metabolic pathways available in horses affected by AM. This genus has interesting features to thrive in the gut as (1) complex systems to sense and adapt to available nutrients, (2) systems to get rid of toxic substances, and (3) the ability to control the environment by interacting with the host's immune system, which is an opportunity to control of other pathogens considered as competitors by *Bacteroides* [78]. Moreover, the relative abundance of *Prevotella* is inversely correlated with that of the *Bacteroides* in human gut microbiota [80], and this is what is observed in the faecal microbiota analysis between CONTROL and AM-NS horses.

The genus *Treponema* is associated with the degradation of structural carbohydrates and correlates positively with different dietary fibres in pigs and horses [66,81]. In this study, *Treponema* was present in faecal microbiota in CONTROL horses and significantly decreased in AM-NS, which could also be explained by the modification of energetic metabolism in AM horses.

The explanations provided remain hypothetical. At this stage, it is not possible to draw definitive conclusions about the role of these bacterial populations in the context of AM based solely on relative abundance. By definition, the analysis of relative abundance reveals populations that are increasing and others that are decreasing relative to each other. This phenomenon is illustrated by the inverse correlation observed between *Prevotella* and *Bacteroides* in humans [80]. This illustration of the ecological niche can also explain variations in populations of genera like *Campylobacter*, *Endomicrobium*, and *Streptococcus*. It is important to remain cautious when interpreting increases or decreases in bacterial populations expressed in relative abundance values: only an absolute quantification of these populations would make it possible to truly quantify the increases and decreases (for example, via RT-PCR). Furthermore, the bacteria present in the gut microbiota are also influenced by the host's metabolism. As a result, observations made in vivo, at the level of the faeces, may not necessarily reflect changes in the microbiota itself following exposure to the protoxins studied. Finally, it is possible that bacteria identified as significantly different between groups are involved in host-related metabolic functions rather than in the direct metabolism of HGA. The discovery of the subclinical character of metabolic processes in horses that have ingested toxins involved in AM [16] suggests that the gut microbiota also has time to modify before the appearance of the clinical phase, which remains acute.

The enzymes responsible for HGA metabolism (i.e., BCAT and BCKDHc) are involved in the catabolism of BCAAs in bacteria [82], leading to the hypothesis that bacteria can transform HGA into toxic metabolic compounds. In mammals, the BCAT exists in two isoenzymes: a mitochondrial (BCATm) ubiquitously present (mainly in skeletal muscle, brain, kidney, and intestine in humans), and a cytosolic (BCATc) isoenzyme, mainly presents in the brain in humans and rats, and in ovary and placenta in rats [83–90]. In contrast to the mitochondrial and cytosolic isoenzymes found in higher eukaryotes, a single form of BCAT is ubiquitously expressed in bacteria. Moreover, this BCAT is involved in the final step of anabolism and in the first step of catabolism of BCAAs, which explains that many bacterial species possess this BCAT. Finally, bacterial BCAT is also distinguished from eukaryotes by broad substrate specificity [91–93]. The mammals BCKDHc is a multienzyme complex of three separate subunits located on the inner surface of the inner mitochondrial membrane [89,94]. The bacterial BCKDHc seems to mainly consist of three subunits, as described for mammalian species [82,95] and is also found in several bacterial species [25,27,28,31,96,97]. Among the bacterial genera highlighted by Aldex function and/or Deseq2 analysis, some of them have in their known genome content one or more copies of the orthologous sequences encoding for the BCAT and/or subunits of BCKDHc (i.e., enzymes involved in the metabolisation of HGA). However, no bacterial genus appears to possess all the expected enzymes, and consequently, the question of the ability of these genera to completely transform HGA in MCPA-linked to carnitine or glycine arises. Interestingly, (1) a pattern of distribution can be observed with the same ortholog missing in groups of genera owning 5/6, 4/6 or 3/6 orthologous sequences and (2) if we compare the pattern of distribution of genera identified in the scientific literature as having the BCKDHc (i.e., *Bacillus*, *Pseudomonas*, *Staphylococcus* [24–26,28–31,96], this pattern is exactly the same as the pattern encountered for the genera having 5/6 orthologous sequences with only the K11381 missing (i.e., 2-oxoisovalerate dehydrogenase E1 component). Unfortunately, these OTUs presented this interesting pattern belonging to the non-clearly identified genera *Bacilli_ge* and *Bacteria_ge*. Indeed, the identification of these OTUs is stopped at the *Bacilli* class and Bacteria kingdom, respectively, preventing further possible characterisation and explanations. Nevertheless, *Bacteria_ge* was also highlighted by Aldex function (i.e., having a significant impact on the global system) and Deseq2 function (i.e., being significantly higher in CoG, AM-S and AM-NS horses vs. CONTROL horses). It will be valuable to investigate these specific OTUs from *Bacteria_ge* and compare them across groups to determine whether these specific OTUs are significantly different. The *Bacilli_ge* was highlighted by the Deseq2 function and was significantly different between CONTROL and AM-S and AM-NS but in a decreasing way from CONTROL to AM-NS: the opposite of the *Bacteria_ge*. This last fact seems to be contradictory (i.e., having the machinery to metabolise HGA but presenting an opposite evolution between the groups). However, some bacteria might possess analogous (non-orthologous) enzymes capable of performing similar functions.

The "Age" parameter did not seem significant in the horse population of the present study, contrary to recent findings in the scientific literature, which reported that diseased horses were either younger than 2 years or older than 10 years, while 90% of CoG were between 2 and 10 years of age [16]. In our study, there was an age overlap between groups. Regarding the "Sex" parameter, Renaud and collaborators found that, among horses exposed to the protoxins, geldings were less susceptible to developing AM compared to intact males and females [16]. However, in our study, neither "Sex" nor the parameters "Health Status" or "Final Outcome" were significantly associated. This may be explained by the lower number of individuals in our study as compared to the above-mentioned one.

As found previously [16], the mean serum concentrations of HGA and MCPA-carnitine were significantly different when comparing CoG and diseased horses. Among the latter, only MCPA-carnitine could differentiate AM-S from AM-NS. Interestingly, MCPA-carnitine was identified as one of the most important blood variables to explain the variation in microbial composition between groups and seemed to correlate better with AM-NS vs. AM-S.

Acylcarnitines are esters formed by the conjugation of acyl groups (notably fatty acids) with carnitine. They are typically divided into four groups: short-chain (C2–C5), medium-chain (C6–C12), long-chain (C13–C20) and very-long-chain (>C21) acylcarnitines. The primary biological function of acylcarnitines is to facilitate the transport of acyl groups from the cytosol into the mitochondrial matrix for β-oxidation, which in turn contributes to cellular energy production. Atypical myopathy is characterised by a severe alteration of the serum acylcarnitines profile with an increase in nearly all acylcarnitines, regardless of their chain length [4,11,13–16,63,98,99]. In the present study, acylcarnitines profiling revealed significant changes comparable to those previously described in the literature. Moreover, C5-carnitine has been highlighted as the best candidate for helping in both the diagnosis and prognosis of the disease. According to the model described by Renaud et al. (2024), a serum concentration of C5-carnitine lower than 3.04 µmol/L would make it possible to identify a diseased horse vs. a cograzer in more than 90% of cases. A second threshold at 12.21 µmol/L of C5-carnitine would identify an animal likely to die (i.e., negative predictive value) in 76% of cases and a surviving animal (i.e., positive predictive value) in 81% of cases [16]. In this study, the analysis of the C5 concentration identified one horse in the AM-S group (i.e., AM-S 04) with a profile similar to that of a CoG, as well as four other cases with prognostic survival estimates that differed from the real outcome: nevertheless, these results are consistent with the percentages announced in the literature. Of the four diseased horses that ultimately died despite having C5 levels indicative of a positive outcome, it should be noted that 3 of them were euthanised following the worsening of clinical signs despite the intensive treatment put in place, and the last was also euthanised, but the reason (i.e., medical or financial constraints) was not clearly identified. Moreover, to the authors' knowledge, none of the CoG developed symptoms of AM during the sampling season, supporting the results obtained from the analysis of these horses' blood concentration of C5 carnitine. Rapid access to C5 assay results remains a significant challenge, preventing clinicians from promptly updating prognoses in cases characterised by severe clinical signs.

Lastly, the C14:1 was identified as one important blood variable to explain the variation in faecal microbiota between groups. The elevated C14:1 level in neonates may indicate a very-long-chain acyl-coenzyme A dehydrogenase deficiency known under the following abbreviation VLCADD, an autosomal recessive disease [100]. In horses, AM is recognised as an acquired multiple acyl-CoA dehydrogenase deficiency [14,63,98]. The long-chain carnitine C14:1 concentration was above the reference range in AM horses [3,13], though serum concentrations showed no significant difference between surviving and deceased horses [15]. In a recent study involving a larger group of AM-affected horses, C14:1 was also identified as one of the variables with the most significant impact on distinguishing groups (comparable to those used to compare faecal microbiota in this paper), further corroborating the results presented here. However, it did not prove to be a reliable diagnostic or prognostic indicator [16]. In humans, tetradecenoylcarnitine (i.e., C14:1), as other long-chain acylcarnitines, plays a role in insulin resistance and in the development of cardiovascular diseases [101]. Most horses affected by AM show elevated plasma concentrations of cardiac troponin I, a specific biomarker of myocardial injury [36,102], and some horses exhibit specific alterations in electrocardiogram (ECG) recordings and cardiac ultrasound examination similar to those observed in mice and humans with VLCADD [103]. Its role in muscle insulin resistance [104] might contribute to explaining the hyperglycaemia observed in AM horses.

In equids, individual variations in faecal microbiota are influenced by factors such as nutrition, management practices, seasonal variation, medications, animal-related factors, pathological conditions, and stress-related factors [105]. To reduce these individual variations, horses included in the present study were all pasturing for a minimum of 6 h per day during a high-risk season for AM (i.e., autumn and spring), including the CONTROL horses that were sampled in November 2020, as 94% of "autumnal" cases occurred between October and December [106]. Seasonal variation and associated weather conditions are known to influence gut microbiota composition in horses [107–109]. This seasonal effect could be attributed to changes in the composition of environmental bacteria (e.g., soil and grass/haylage microbiota) or in the nutrient composition of pasture which are in turn influenced by climatic conditions [107,109,110]. In this study, AM-S and AM-NS horses were sampled in autumn and spring, while CoG and CONTROL horses were sampled in autumn. However, seasonal variations between winter and summer are reported to have only a minor impact on faecal microbiota, accounting for 2.8% of the variation, according to Theelen et al. (2021) [108]. Similarly, the variations between autumn and spring in the present study could also be considered minor. It is important to note that the year of sampling differed between groups (2016, 2017, and 2018 for clinical cases of AM, 2018 for selected CoG, and 2020 for CONTROL horses), which may introduce potential biases. To minimise the impact of this temporal variation, samples were collected under a strict protocol, stored immediately in a conservative medium, and frozen at −20 °C until analysis [42,111,112]. Moreover, analyses were conducted on a regular basis over the years to ensure that samples did not remain frozen for extended periods. Finally, sampling was also standardised using a protocol described by Stewart et al. (2018) [42], further minimising potential sampling bias.

There are some limitations in this study. Some of them are already described by Wimmer-Scherr et al. (2021): the potential bias introduced by euthanasia for ethical reasons based on the severity of the clinical signs in the AM-NS group as well as the possible medications received by some horses prior to referral to the clinic [35]. The method used for microbiota assessment also presents inherent limitations, potentially favouring or underestimating certain bacterial taxa due to the lack of absolute quantification and selection bias during the process (e.g., DNA extraction, primer selection, PCR amplification, and bioinformatics parameters) [113].

When a modification of the intestinal microbiota is associated with a pathology, the question arises whether this modification reflects the state of health of the host or influences the host's health. Recently, Renaud et al. (2022) suggested that protoxins may be transformed by rumen microbiota, particularly in species with a long retention time, which would protect these species from developing AM-clinical signs. Indeed, the hypothesis proposed is that having a proximal fermentation compartment located before the absorption site of amino acids might be protective while having a distal fermentation compartment located after the absorption site of protoxins—as is the case in horses—might make the species more sensitive to poisoning [40]. This hypothesis can be partly explained by the fact that (1) certain bacteria can metabolise peptides and amino acids, and the protoxins are non-proteinogenic amino acids [114,115], and (2) some bacteria possess the enzymes involved in the metabolism of BCAAs [6,114], HGA, and MCPrG [6].

The identification of bacteria within the equine microbiota that can metabolise HGA and MCPrG into their corresponding toxic metabolites or potentially degrade these protoxins could be valuable in understanding species sensitivity. Moreover, this could contribute to a broader comprehension of intoxication and potentially aid in the discovery of molecules and/or bacteria capable of preventing this intoxication. To study the intestinal microbiota in this context, the use of alternative models, such as in vitro dynamic (i.e., SHIME® or static fermentation models (i.e., batch), makes sense [116–118]. In addition to aligning with the three R's—replacement, reduction, and refinement—this type of model would eliminate the direct influence of host metabolism on the microbiota by using faeces of healthy donors (i.e., CONTROL horses) and by adding the studied challenge (for example, the addition of HGA). As such, the changes observed in the digestive microbiota would be directly linked to the challenge applied to the system: the addition of protoxins in the AM model. The identification of bacteria that could play a role in HGA and/or MCPrG poisoning would then be easier. Another benefit of this kind of in vitro dynamic or static fermentation model is that it can offer opportunities for studying treatments by also adding targeted molecules. Finally, another possibility would be to use metagenomic shotgun analysis. This analysis makes it possible to describe the taxonomic composition of a community of organisms and its diversity, as well as its genes and, therefore, its functional capacities.

5. Conclusions

For the first time, a correlation has been observed between blood parameters and the intestinal microbiota of horses suffering from AM. At this stage, it remains to be determined whether these changes result directly from the protoxins' effect on the microbiota, the metabolism of protoxins by bacteria, and/or the host's pathological state. Further investigation is necessary to elucidate the underlying mechanisms and determine the specific contributions of each factor. Understanding these relationships could deepen our knowledge of the role of the intestinal microbiota in AM and open the path to potential therapeutic strategies.

Author Contributions: Conceptualisation, A.-C.F. and C.C.; methodology, A.-C.F. and B.T.; validation, A.-C.F. and B.T.; formal analysis, A.-C.F., B.T. and F.B.; investigation, A.-C.F., C.C., B.R., C.-J.K., G.v.L. and K.P.; resources, D.-M.V., F.B., P.G. and B.T.; data curation, B.T.; writing—original draft preparation, A.-C.F.; writing—review and editing, A.-C.F., C.C., D.-M.V., B.T., B.R., G.v.L., L.L., C.P.W., G.D. and C.-J.K.; visualisation, A.-C.F. and B.T.; supervision, D.-M.V. and P.G.; project administration, A.-C.F.; funding acquisition, D.-M.V. All authors have read and agreed to the published version of the manuscript.

Funding: This research was funded by Wallonie Agriculture SPW (Service Public de Wallonie, Belgium) Grants No. D31-1381/S1-SAMA and D65-1418-SAMA II.

Institutional Review Board Statement: Ethical review and approval were waived for this study because all procedures in this study adhered to both national and international guidelines on animal welfare. The Animal Ethics Committee of the University of Liege was consulted, and it was confirmed that the sampling process was part of routine veterinary practice for diagnosing or preventing atypical myopathy.

Informed Consent Statement: Informed consent was obtained from all subjects involved in the study.

Data Availability Statement: Raw amplicon sequencing libraries were submitted to the NCBI database under bioproject number PRJNA1170059.

Acknowledgments: The authors thank Benjamin Klein for technical support.

Conflicts of Interest: The authors declare no conflicts of interest.

Appendix A

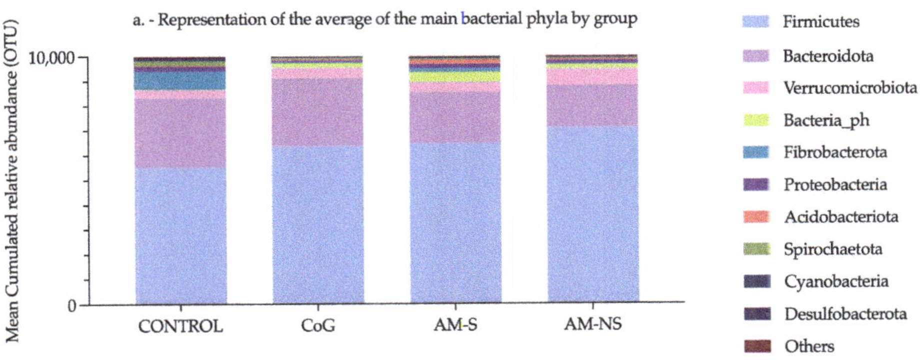

Figure A1. *Cont.*

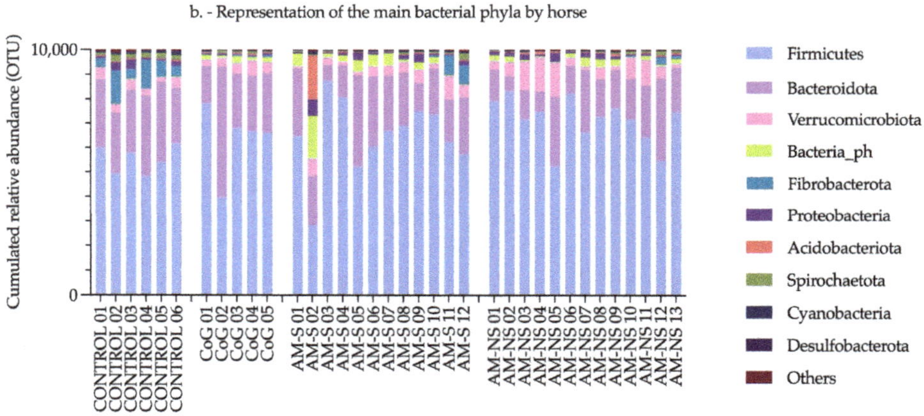

Figure A1. Main dominant bacterial phyla: (**a**) by group and (**b**) by horse. The horses represented are control horses (CONTROL), cograzers (CoG), survivors (AM-S) and non-survivors (AM-NS) of atypical myopathy.

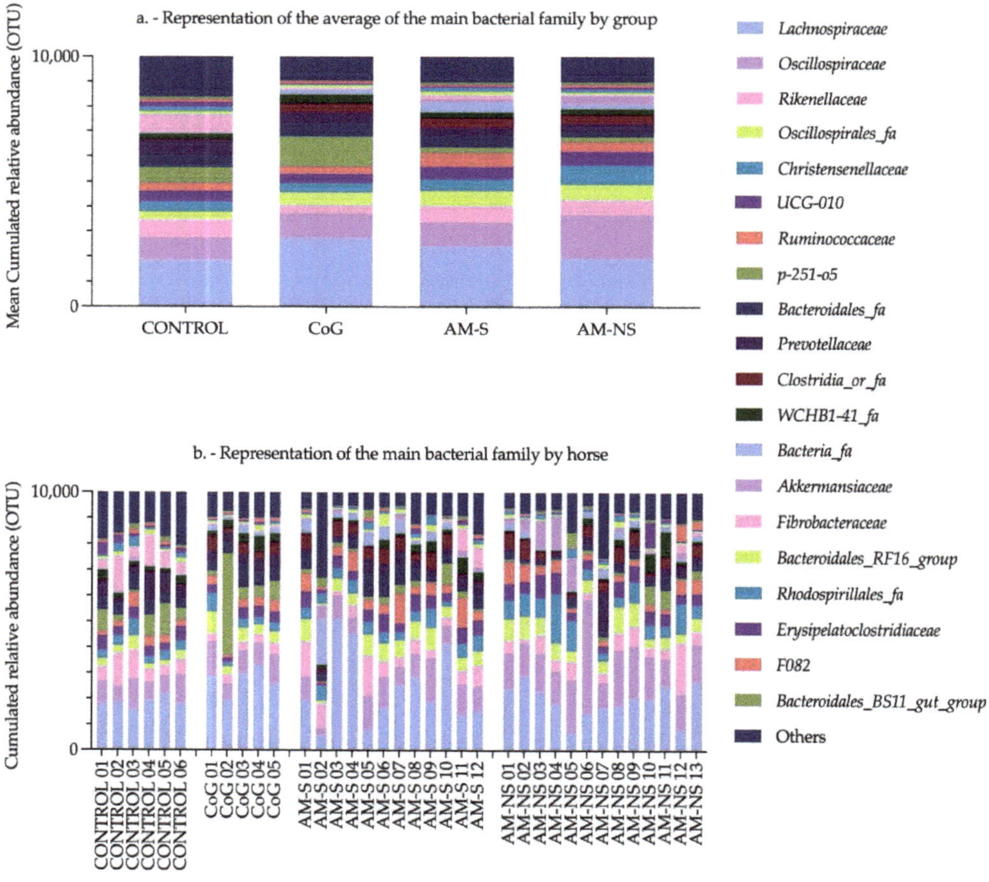

Figure A2. Main dominant bacterial family: (**a**) by group and (**b**) by horse. The horses represented are control horses (CONTROL), cograzers (CoG), survivors (AM-S) and non-survivors (AM-NS) of atypical myopathy.

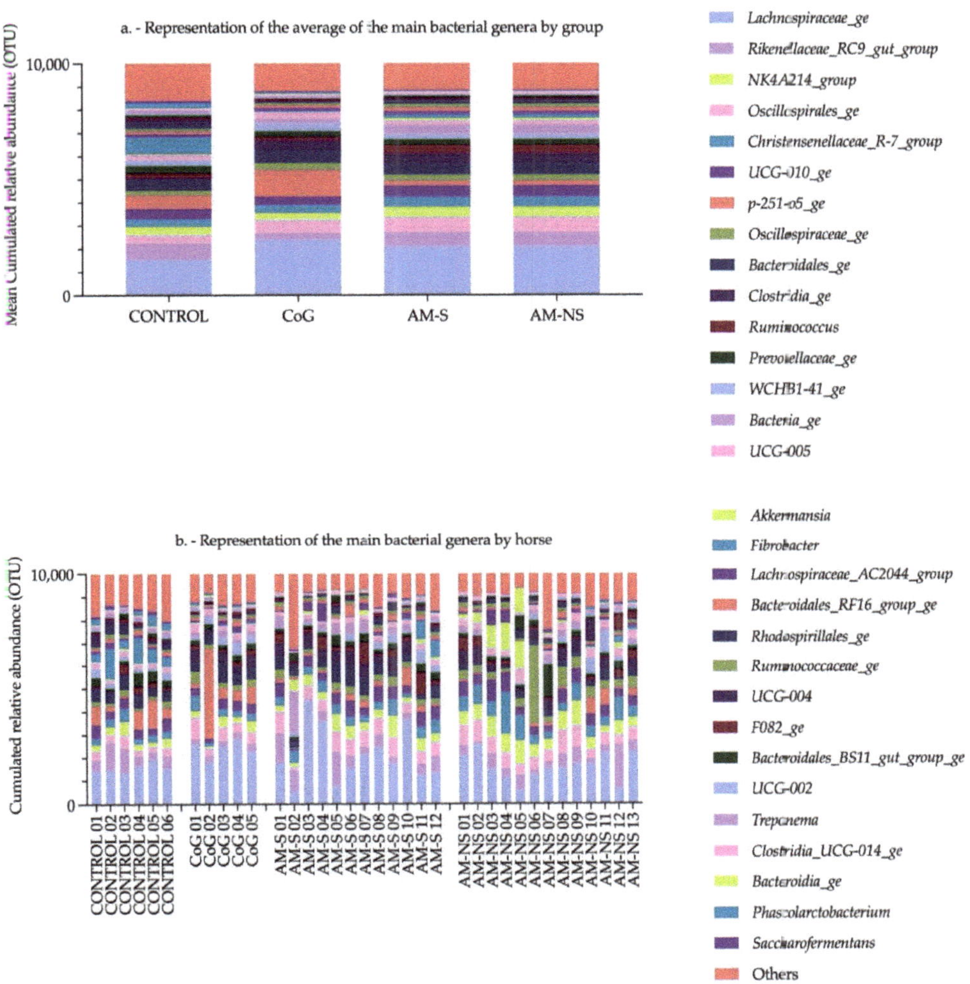

Figure A3. Main dominant bacterial genera: (**a**) by group and (**b**) by horse. The horses represented are control horses (CONTROL), cograzers (CoG), survivors (AM-S) and non-survivors (AM-NS) of atypical myopathy.

Table A1. Serum concentrations of hypoglycin A (µmol/L) and MCPA-carnitine (nmol/L).

		CONTROL	CoG	AM-S	AM-NS	Diseased Horses
HGA	Mean ± SD CI [LLCI-ULCI]	<LD	1.46 ± 1.81 [a] [0.00–3.05]	4.59 ± 3.67 [2.52–6.67]	7.32 ± 5.61 [4.28–10.37]	6.01 ± 4.88 [a] [3.88–7.72]
MCPA-carnitine	Mean ± SD CI [LLCI-ULCI]	<LD	9.63 ± 14.03 [b] [0.00–21.92]	369.62 ± 972.82 [c] [0.00–920.04]	507.73 ± 485.45 [c] [243.84–771.62]	441.44 ± 746.02 [b] [149–733.88]

HGA: Hypoglycin A/MCPA-carnitine: methylenecyclopropylacetyl-carnitine/< LD = below the limit of detection (0.09 µmol/L for HGA and 0.01 nmol/L for MCPA-carnitine), SD = standard deviation, CI = Confidence interval, LLCI = Lower Limit of Confidence Interval, ULCI = Upper Limit of Confidence Interval. The groups represented are control horses (CONTROL), cograzers (CoG), survivors (AM-S) and non-survivors (AM-NS) of atypical myopathy and diseased horses. a—Significant difference in mean serum concentration of HGA, $p < 0.01$. b—Significant difference in mean serum concentration of MCPA-carnitine, $p < 0.0001$. c—Significant difference in mean serum concentration of MCPA-carnitine, $p < 0.05$.

Table A2. Serum concentrations of selected acylcarnitines (µmol/L).

		CONTROL	CoG	AM-S	AM-NS	Diseased Horses
C2	Mean ± SD CI [LLCI-ULCI]	7.43 ± 0.96 [6.67–8.39]	6.59 ± 2.15 [4.71–8.73]	23.87 ± 14.64 * [15.58–38.51]	43.05 ± 25.42 * [29.23–68.47]	33.84 ± 22.74 * [24.93–56.53]
C4	Mean ± SD CI [LLCI-ULCI]	0.65 ± 0.28 [0.43–0.93]	0.85 ± 0.43 [0.48–1.28]	10.76 ± 7.02 * [6.79–17.78]	21.06 ± 15.16 * [12.81–36.22]	16.12 ± 12.85 * [11.08–28.97]
C5	Mean ± SD CI [LLCI-ULCI]	0.29 ± 0.05 [0.25–0.33]	0.93 ± 0.51 * [0.48–1.44]	9.52 ± 7.83 * [5.10–17.35]	21.64 ± 14.15 * [13.94–35.79]	15.82 ± 12.90 * [10.77–28.72]
C10	Mean ± SD CI [LLCI-ULCI]	0.01 ± 0.00 [0.01–0.01]	0.04 ± 0.03 * [0.01–0.07]	0.55 ± 0.54 * [0.25–1.10]	0.96 ± 1.39 * [0.21–2.35]	0.77 ± 1.07 * [0.35–1.84]
C12:1	Mean ± SD CI [LLCI-ULCI]	0.01 ± 0.00 [0.00–0.01]	0.01 ± 0.01 [0.00–0.02]	0.20 ± 0.13 * [0.13–0.33]	0.28 ± 0.26 * [0.14–0.55]	0.24 ± 0.21 * [0.16–0.45]
C14	Mean ± SD CI [LLCI-ULCI]	0.01 ± 0.00 [0.01–0.02]	0.05 ± 0.07 [0.00–0.12]	0.22 ± 0.13 * [0.14–0.35]	0.34 ± 0.28 * [0.19–0.62]	0.28 ± 0.23 * [0.19–0.51]
C14:1	Mean ± SD CI [LLCI-ULCI]	0.01 ± 0.00 [0.01–0.01]	0.03 ± 0.01 [0.01–0.04]	0.33 ± 0.22 * [0.21–0.55]	0.54 ± 0.56 * [0.24–1.10]	0.44 ± 0.43 * [0.27–0.87]
C18:1	Mean ± SD CI [LLCI-ULCI]	0.04 ± 0.02 [0.03–0.06]	0.08 ± 0.06 [0.02–0.14]	0.61 ± 0.48 * [0.34–1.09]	0.92 ± 0.86 * [0.46–1.78]	0.77 ± 0.71 * [0.49–1.48]

* Mean over the percentile 99 of reference range obtained with control horses, SD = standard deviation, CI = Confidence interval, LLCI = Lower Limit of Confidence Interval, ULCI = Upper Limit of Confidence Interval. The groups represented are control horses (CONTROL), cograzers (CoG), survivors (AM-S) and non-survivors (AM-NS) of atypical myopathy and diseased horses.

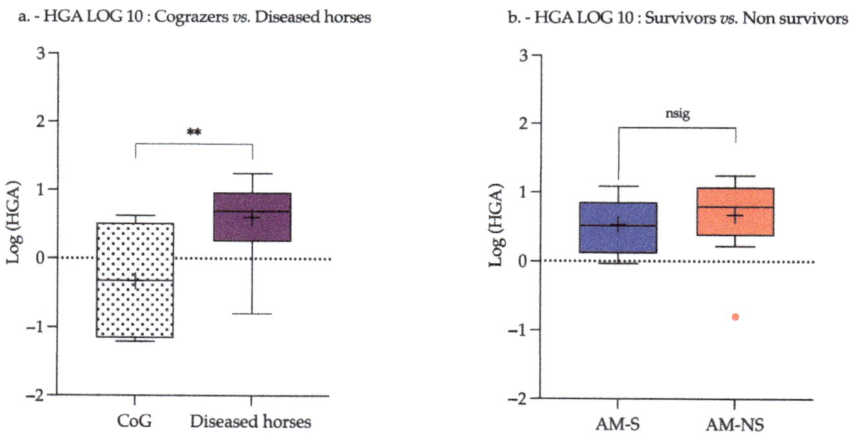

Figure A4. *Cont.*

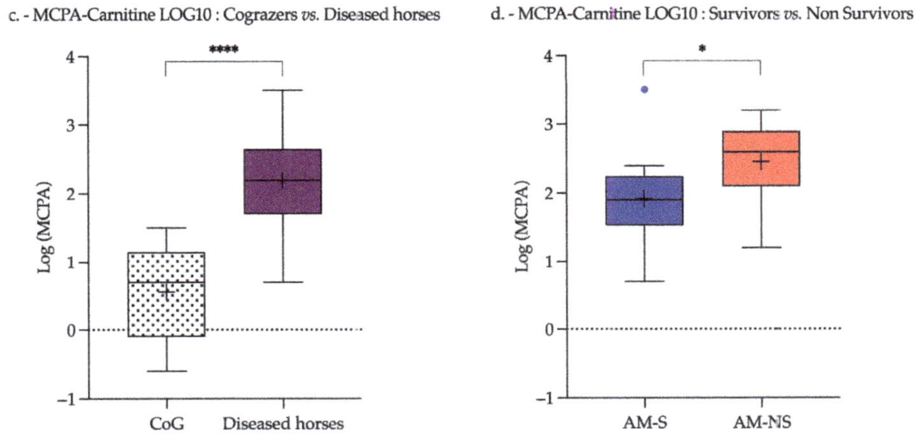

Figure A4. Serum hypoglycin A (HGA) (**a**,**b**) and methylenecyclopropylacetic-carnitine (MCPA-carnitine) (**c**,**d**) concentrations. Boxes range from the 25th to 75th percentiles. Means are represented by horizontal lines and medians by a cross. Box plot whiskers were established using the Tukey method. The groups are control horses (CONTROL), cograzers (CoG), survivors (AM-S) and non-survivors (AM-NS) of atypical myopathy and diseased horses. Significantly different with a p-value of 0.05 or less: * < 0.05; ** < 0.01; **** < 0.0001, Not significant "nsig".

References

1. Van Galen, G.; Marcillaud Pitel, C.; Saegerman, C.; Patarin, F.; Amory, H.; Baily, J.D.; Cassart, D.; Gerber, V.; Hahn, C.; Harris, P.; et al. European Outbreaks of Atypical Myopathy in Grazing Equids (2006–2009): Spatiotemporal Distribution, History and Clinical Features. *Equine Vet. J.* **2012**, *44*, 614–620. [CrossRef] [PubMed]
2. Valberg, S.J.; Sponseller, B.T.; Hegeman, A.D.; Earing, J.; Bender, J.B.; Martinson, K.L.; Patterson, S.E.; Sweetman, L. Seasonal Pasture Myopathy/Atypical Myopathy in North America Associated with Ingestion of Hypoglycin A within Seeds of the Box Elder Tree. *Equine Vet. J.* **2013**, *45*, 419–426. [CrossRef] [PubMed]
3. Votion, D.M.; van Galen, G.; Sweetman, L.; Boemer, F.; de Tullio, P.; Dopagne, C.; Lefère, L.; Mouithys-Mickalad, A.; Patarin, F.; Rouxhet, S.; et al. Identification of Methylenecyclopropyl Acetic Acid in Serum of European Horses with Atypical Myopathy. *Equine Vet. J.* **2014**, *46*, 146–149. [CrossRef] [PubMed]
4. Bochnia, M.; Sander, J.; Ziegler, J.; Terhardt, M.; Sander, S.; Janzen, N.; Cavalleri, J.M.V.; Zuraw, A.; Wensch-Dorendorf, M.; Zeyner, A. Detection of MCPG Metabolites in Horses with Atypical Myopathy. *PLoS ONE* **2019**, *14*, e0211698. [CrossRef] [PubMed]
5. von Holt, C.; Chang, J.; von Holt, M.; Bohm, H. Metabolism and Metabolic Effects of Hypoglycin. *Biochim. Biophys. Acta* **1964**, *90*, 611–613. [CrossRef] [PubMed]
6. Melde, K.; Jackson, S.; Bartlett, K.; Stanley, H.; Sherratt, A.; Ghisla, S. Metabolic Consequences of Methylenecyclopropylglycine Poisoning in Rats. *Biochem. J.* **1991**, *274*, 395–400. [CrossRef]
7. Ichihara, A.; Koyama, E. Transaminase of Branched Chain Amino Acids: I. Branched Chain Amino Acids-α-Ketoglutarate Transaminase. *J. Biochem.* **1966**, *59*, 160–169. [CrossRef] [PubMed]
8. Danner, D.J.; Lemmon, S.K.; Besharse, J.C.; Elsas, L.J. Purification and Characterization of Branched Chain Alpha-Ketoacid Dehydrogenase from Bovine Liver Mitochondria. *J. Biol. Chem.* **1979**, *254*, 5522–5526. [CrossRef]
9. Sander, J.; Terhardt, M.; Janzen, N.; Renaud, B.; Kruse, C.J.; François, A.C.; Wouters, C.P.; Boemer, F.; Votion, D.M. Tissue Specific Distribution and Activation of Sapindaceae Toxins in Horses Suffering from Atypical Myopathy. *Animals* **2023**, *13*, 2410. [CrossRef]
10. Bochnia, M.; Scheidemann, W.; Ziegler, J.; Sander, J.; Vollstedt, S.; Glatter, M.; Janzen, N.; Terhardt, M.; Zeyner, A. Predictive Value of Hypoglycin A and Methylencyclopropylacetic Acid Conjugates in a Horse with Atypical Myopathy in Comparison to Its Cograzing Partners. *Equine Vet. Educ.* **2016**, *30*, 24–28. [CrossRef]
11. Karlíková, R.; Široká, J.; Jahn, P.; Friedecký, D.; Gardlo, A.; Janečková, H.; Hrdinová, F.; Drábková, Z.; Adam, T. Equine Atypical Myopathy: A Metabolic Study. *Vet. J.* **2016**, *216*, 125–132. [CrossRef] [PubMed]
12. Von Holt, C.; Von Holt, M.; Böhm, H. Metabolic Effects of Hypoglycin and Methylenecyclopropaneacetic Acid. *Biochim. Biophys. Acta (BBA) Lipids Lipid Metab.* **1966**, *125*, 11–21. [CrossRef]
13. Lemieux, H.; Boemer, F.; van Galen, G.; Serteyn, D.; Amory, H.; Baise, E.; Cassart, D.; van Loon, G.; Marcillaud-Pitel, C.; Votion, D.M. Mitochondrial Function Is Altered in Horse Atypical Myopathy. *Mitochondrion* **2016**, *30*, 35–41. [CrossRef]

14. Westermann, C.M.; Dorland, L.; Votion, D.M.; de Sain-van der Velden, M.G.M.; Wijnberg, I.D.; Wanders, R.J.A.; Spliet, W.G.M. Testerink, N.; Berger, R.; Ruiter, J.P.N.; et al. Acquired Multiple Acyl-CoA Dehydrogenase Deficiency in 10 Horses with Atypical Myopathy. *Neuromuscul. Disord.* **2008**, *18*, 355–364. [CrossRef]
15. Boemer, F.; Detilleux, J.; Cello, C.; Amory, H.; Marcillaud-Pitel, C.; Richard, E.; Van Galen, G.; Van Loon, G.; Lefère, L.; Votion D.M. Acylcarnitines Profile Best Predicts Survival in Horses with Atypical Myopathy. *PLoS ONE* **2017**, *12*, e182761. [CrossRef]
16. Renaud, B.; Kruse, C.-J.; François, A.-C.; Cesarini, C.; van Loon, G.; Palmers, K.; Boemer, F.; Luis, G.; Gustin, P.; Votion, D.-M. Large-Scale Study of Blood Markers in Equine Atypical Myopathy Reveals Subclinical Poisoning and Advances in Diagnostic and Prognostic Criteria. *Environ. Toxicol. Pharmacol.* **2024**, 104515. [CrossRef]
17. Boemer, F.; Deberg, M.; Schoos, R.; Baise, E.; Amory, H.; Gault, G.; Carlier, J.; Gaillard, Y.; Marcillaud-Pitel, C.; Votion, D. Quantification of Hypoglycin A in Serum Using ATRAQ® Assay. *J. Chromatogr. B Analyt Technol. Biomed. Life Sci.* **2015**, *997*, 75–80. [CrossRef] [PubMed]
18. Mathis, D.; Sass, J.O.; Graubner, C.; Schoster, A. Diagnosis of Atypical Myopathy Based on Organic Acid and Acylcarnitine Profiles and Evolution of Biomarkers in Surviving Horses. *Mol. Genet. Metab. Rep.* **2021**, *29*, 100827. [CrossRef] [PubMed]
19. Votion, D.M. Analysing Hypoglycin A, Methylenecyclopropylacetic Acid Conjugates and Acylcarnitines in Blood to Confirm the Diagnosis and Improve Our Understanding of Atypical Myopathy. *Equine Vet. Educ.* **2018**, *30*, 29–30. [CrossRef]
20. Koike, M.; Koike, K. Structure, Assembly and Function of Mammalian Alpha-Keto Acid Dehydrogenase Complexes. *Adv. Biophys* **1976**, *9*, 187–227.
21. Yvon, M.; Chambellon, E.; Bolotin, A.; Roudot-Algaron, F. Characterization and Role of the Branched-Chain Aminotransferase (BcaT) Isolated from Lactococcus Lactis Subsp. Cremoris NCDO 763. *Appl. Environ. Microbiol.* **2000**, *66*, 571. [CrossRef] [PubMed]
22. Yu, X.; Wang, X.; Engel, P.C. The Specificity and Kinetic Mechanism of Branched-Chain Amino Acid Aminotransferase from Escherichia Coli Studied with a New Improved Coupled Assay Procedure and the Enzyme's Potential for Biocatalysis. *FEBS J.* **2014**, *281*, 391–400. [CrossRef]
23. Namba, Y.; Yoshizawa, K.; Ejima, A.; Hayashi, T.; Kaneda, T. Coenzyme A- and Nicotinamide Adenine Dinucleotide-Dependent Branched Chain α-Keto Acid Dehydrogenase: I. PURIFICATION AND PROPERTIES OF THE ENZYME FROM BACILLUS SUBTILIS. *J. Biol. Chem.* **1969**, *244*, 4437–4447. [CrossRef]
24. Willecke, K.; Pardee, A.B. Fatty Acid-Requiring Mutant of Bacillus Subtilis Defective in Branched Chain α-Keto Acid Dehydrogenase. *J. Biol. Chem.* **1971**, *246*, 5264–5272. [CrossRef]
25. Wang, G.-F.; Kuriki, T.; Roy, K.L.; Kaneda, T. The Primary Structure of Branched-Chain α-Oxo Acid Dehydrogenase from Bacillus Subtilis and Its Similarity to Other α-Oxo Acid Dehydrogenases. *Eur. J. Biochem.* **1993**, *213*, 1091–1099. [CrossRef] [PubMed]
26. Oku, H.; Kaneda, T. Biosynthesis of Branched-Chain Fatty Acids in Bacillus Subtilis. A Decarboxylase Is Essential for Branched-Chain Fatty Acid Synthetase. *J. Biol. Chem.* **1988**, *263*, 18386–18396. [CrossRef] [PubMed]
27. Martin, R.R.; Marshall, V.D.; Sokatch, J.R.; Unger, L. Common Enzymes of Branched-Chain Amino Acid Catabolism in Pseudomonas Putida. *J. Bacteriol.* **1973**, *115*, 198. [CrossRef]
28. Sokatch, J.R.; McCully, V.; Roberts, C.M. Purification of a Branched-Chain Keto Acid Dehydrogenase from Pseudomonas Putida. *J. Bacteriol.* **1981**, *148*, 647–652. [CrossRef]
29. Sykes, P.J.; Menard, J.; McCully, V.; Sokatch, J.R. Conjugative Mapping of Pyruvate, 2-Ketoglutarate, and Branched-Chain Keto Acid Dehydrogenase Genes in Pseudomonas Putida Mutants. *J. Bacteriol.* **1985**, *162*, 203. [CrossRef]
30. Singh, V.K.; Hattangady, D.S.; Giotis, E.S.; Singh, A.K.; Chamberlain, N.R.; Stuart, M.K.; Wilkinson, B.J. Insertional Inactivation of Branched-Chain α-Keto Acid Dehydrogenase in Staphylococcus Aureus Leads to Decreased Branched-Chain Membrane Fatty Acid Content and Increased Susceptibility to Certain Stresses. *Appl. Environ. Microbiol.* **2008**, *74*, 5882–5890. [CrossRef] [PubMed]
31. Singh, V.K.; Sirobhushanam, S.; Ring, R.P.; Singh, S.; Gatto, C.; Wilkinson, B.J. Roles of Pyruvate Dehydrogenase and Branched-Chain α-Keto Acid Dehydrogenase in Branched-Chain Membrane Fatty Acid Levels and Associated Functions in Staphylococcus Aureus. *J. Med. Microbiol.* **2018**, *67*, 570. [CrossRef] [PubMed]
32. Dougal, K.; de la Fuente, G.; Harris, P.A.; Girdwood, S.E.; Pinloche, E.; Newbold, C.J. Identification of a Core Bacterial Community within the Large Intestine of the Horse. *PLoS ONE* **2013**, *8*, e077660. [CrossRef] [PubMed]
33. Costa, M.C.; Silva, G.; Ramos, R.V.; Staempfli, H.R.; Arroyo, L.G.; Kim, P.; Weese, J.S. Characterization and Comparison of the Bacterial Microbiota in Different Gastrointestinal Tract Compartments in Horses. *Vet. J.* **2015**, *205*, 74–80. [CrossRef] [PubMed]
34. Ericsson, A.C.; Johnson, P.J.; Lopes, M.A.; Perry, S.C.; Lanter, H.R. A Microbiological Map of the Healthy Equine Gastrointestinal Tract. *PLoS ONE* **2016**, *11*, e166523. [CrossRef]
35. Wimmer-Scherr, C.; Taminiau, B.; Renaud, B.; van Loon, G.; Palmers, K.; Votion, D.; Amory, H.; Daube, G.; Cesarini, C. Comparison of Fecal Microbiota of Horses Suffering from Atypical Myopathy and Healthy Co-Grazers. *Animals* **2021**, *11*, 506. [CrossRef] [PubMed]
36. Votion, D.; Linden, A.; Saegerman, C.; Engels, P.; Erpicum, M.; Thiry, E.; Delguste, C.; Rouxhet, S.; Demoulin, V.; Navet, R.; et al. History and Clinical Features of Atypical Myopathy in Horses in Belgium (2000-2005). *J. Vet. Intern. Med.* **2007**, *21*, 1380–1391. [CrossRef] [PubMed]

37. Bochnia, M.; Ziegler, J.; Sander, J.; Uhlig, A.; Schaefer, S.; Vollstedt, S.; Glatter, M.; Abel, S.; Recknagel, S.; Schusser, G.F.; et al. Hypoglycin a Content in Blood and Urine Discriminates Horses with Atypical Myopathy from Clinically Normal Horses Grazing on the Same Pasture. *PLoS ONE* **2015**, *10*, e0136785. [CrossRef] [PubMed]
38. Baise, E.; Habyarimana, J.A.; Amory, H.; Boemer, F.; Douny, C.; Gustin, P.; Marcillaud-Pitel, C.; Patarin, F.; Weber, M.; Votion, D.M. Samaras and Seedlings of Acer Pseudoplatanus Are Potential Sources of Hypoglycin A Intoxication in Atypical Myopathy without Necessarily Inducing Clinical Signs. *Equine Vet. J.* **2016**, *48*, 414–417. [CrossRef]
39. Grönzahl, G.; Berglund, A.; Skidell, J.; Bondesson, U.; Salomonsson, M. Detection of the Toxin Hypoglycin A in Pastured Horses and in the European Sycamore Maple Tree (Acer Pseudoplatanus) During Two Outbreaks of Atypical Myopathy in Sweden. *Equine Vet. J.* **2015**, *47*, 22. [CrossRef]
40. Renaud, B.; Kruse, C.J.; François, A.C.; Grund, L.; Bunert, C.; Brisson, L.; Boemer, F.; Gault, G.; Ghislain, B.; Petitjean, T.; et al. Acer Pseudoplatanus: A Potential Risk of Poisoning for Several Herbivore Species. *Toxins* **2022**, *14*, 512. [CrossRef]
41. Van Galen, G.; Saegerman, C.; Marcillaud Pitel, C.; Patarin, F.; Amory, H.; Baily, J.D.; Cassart, D.; Gerber, V.; Hahn, C.; Harris, P.; et al. European Outbreaks of Atypical Myopathy in Grazing Horses (2006-2009): Determination of Indicators for Risk and Prognostic Factors. *Equine Vet. J.* **2012**, *44*, 621–625. [CrossRef]
42. Stewart, H.L.; Pitta, D.; Indugu, N.; Vecchiarelli, B.; Engiles, J.B.; Southwood, L.L. Characterization of the Fecal Microbiota of Healthy Horses. *Am. J. Vet. Res.* **2018**, *79*, 811–819. [CrossRef]
43. Cerri, S.; Taminiau, B.; de Lusancay, A.H.C.; Lecoq, L.; Amory, H.; Daube, G.; Cesarini, C. Effect of Oral Administration of Omeprazole on the Microbiota of the Gastric Glandular Mucosa and Feces of Healthy Horses. *J. Vet. Intern. Med.* **2020**, *34*, 2727–2737. [CrossRef] [PubMed]
44. Schloss, P.D.; Westcott, S.L.; Ryabin, T.; Hall, J.R.; Hartmann, M.; Hollister, E.B.; Lesniewski, R.A.; Oakley, B.B.; Parks, D.H.; Robinson, C.J.; et al. Introducing Mothur: Open-Source, Platform-Independent, Community-Supported Software for Describing and Comparing Microbial Communities. *Appl. Environ. Microbiol.* **2009**, *75*, 7537–7541. [CrossRef] [PubMed]
45. Rognes, T.; Flouri, T.; Nichols, B.; Quince, C.; Mahé, F. VSEARCH: A Versatile Open Source Tool for Metagenomics. *PeerJ* **2016**, *4*, e2584. [CrossRef] [PubMed]
46. Quast, C.; Pruesse, E.; Yilmaz, P.; Gerken, J.; Schweer, T.; Yarza, P.; Peplies, J.; Glöckner, F.O. The SILVA Ribosomal RNA Gene Database Project: Improved Data Processing and Web-Based Tools. *Nucleic Acids Res.* **2013**, *41*. [CrossRef] [PubMed]
47. Oksanen, J.; Simpso, G.L.; Blanchet, F.G.; Kindt, R.; Legendre, P.; Minchin, P.R.; O'Hara, R.B.; Solymos, P.; Stevens, M.H.M.; Szoecs, E.; et al. Package "vegan"—Community Ecology Package. Available online: https://cran.r-project.org/web/packages/vegan/vegan.pdf (accessed on 9 October 2024).
48. Kers, J.G.; Saccenti, E. The Power of Microbiome Studies: Some Considerations on Which Alpha and Beta Metrics to Use and How to Report Results. *Front. Microbiol.* **2022**, *12*. [CrossRef] [PubMed]
49. Morris, E.K.; Caruso, T.; Buscot, F.; Fischer, M.; Hancock, C.; Maier, T.S.; Meiners, T.; Müller, C.; Obermaier, E.; Prati, D.; et al. Choosing and Using Diversity Indices: Insights for Ecological Applications from the German Biodiversity Exploratories. *Ecol. Evol.* **2014**, *4*, 3514–3524. [CrossRef]
50. Chao, A.; Chazdon, R.L.; Shen, T.J. A New Statistical Approach for Assessing Similarity of Species Composition with Incidence and Abundance Data. *Ecol. Lett.* **2005**, *8*, 148–159. [CrossRef]
51. Oksanen, J.; Kindt, R.; Simpson, G. Vegan3d: Static and Dynamic 3D and Editable Interactive Plots for the "vegan" Package_. R Package Version 1.3-0. Available online: https://cran.r-project.org/web/packages/vegan3d/vegan3d.pdf (accessed on 9 October 2024).
52. Martinez Arbizu, P. PairwiseAdonis: Pairwise Multilevel Comparison Using Adonis_. R Package Version 0.4.1. Available online: https://github.com/pmartinezarbizu/pairwiseAdonis (accessed on 9 October 2024).
53. Fernandes, A.D.; Macklaim, J.M.; Linn, T.G.; Reid, G.; Gloor, G.B. ANOVA-Like Differential Expression (ALDEx) Analysis for Mixed Population RNA-Seq. *PLoS ONE* **2013**, *8*, e067019. [CrossRef]
54. Douglas, G.M.; Maffei, V.J.; Zaneveld, J.R.; Yurgel, S.N.; Brown, J.R.; Taylor, C.M.; Huttenhower, C.; Langille, M.G.I. PICRUSt2 for Prediction of Metagenome Functions. *Nat. Biotechnol.* **2020**, *38*, 685–688. [CrossRef] [PubMed]
55. Chace, D.H.; Pons, R.; Chiriboga, C.A.; McMahon, D.J.; Tein, I.; Naylor, E.W.; De Vivo, D.C. Neonatal Blood Carnitine Concentrations: Normative Data by Electrospray Tandem Mass Spectometry. *Pediatr. Res.* **2003**, *53*, 823–829. [CrossRef] [PubMed]
56. Marcus, W. Beck Ggord: Ordination Plots with Ggplot2. R Package Version 1.1.8. Available online: https://zenodo.org/badge/latestdoi/35334615 (accessed on 9 October 2024).
57. Gagliardi, A.; Totino, V.; Cacciotti, F.; Iebba, V.; Neroni, B.; Bonfiglio, G.; Trancassini, M.; Passariello, C.; Pantanella, F.; Schippa, S. Rebuilding the Gut Microbiota Ecosystem. *Int. J. Environ. Res. Public. Health* **2018**, *15*, 1679. [CrossRef] [PubMed]
58. Lin, C.; Stahl, D.A. Taxon-Specific Probes for the Cellulolytic Genus Fibrobacter Reveal Abundant and Novel Equine-Associated Populations. *Appl. Environ. Microbiol.* **1995**, *61*, 1348–1351. [CrossRef] [PubMed]
59. Elzinga, S.E.; Weese, J.S.; Adams, A.A. Comparison of the Fecal Microbiota in Horses With Equine Metabolic Syndrome and Metabolically Normal Controls Fed a Similar All-Forage Diet. *J. Equine Vet. Sci.* **2016**, *44*, 9–16. [CrossRef]

60. Froidurot, A.; Julliand, V. Cellulolytic Bacteria in the Large Intestine of Mammals. *Gut Microbes* **2022**, *14*, 2031694. [CrossRef]
61. Kobayashi, R.; Nagaoka, K.; Nishimura, N.; Koike, S.; Takahashi, E.; Niimi, K.; Murase, H.; Kinjo, T.; Tsukahara, T.; Inoue, R. Comparison of the Fecal Microbiota of Two Monogastric Herbivorous and Five Omnivorous Mammals. *Anim. Sci. J.* **2020**, *91*, e13366. [CrossRef] [PubMed]
62. Kruse, C.J.; Dieu, M.; Renaud, B.; François, A.C.; Stern, D.; Demazy, C.; Burteau, S.; Boemer, F.; Art, T.; Renard, P.; et al. New Pathophysiological Insights from Serum Proteome Profiling in Equine Atypical Myopathy. *ACS Omega* **2024**, *9*, 6505–6526. [CrossRef]
63. Westermann, C.M.; de Sain-van der Velden, M.G.M.; van der Kolk, J.H.; Berger, R.; Wijnberg, I.D.; Koeman, J.P.; Wanders, R.J.A.; Lenstra, J.A.; Testerink, N.; Vaandrager, A.B.; et al. Equine Biochemical Multiple Acyl-CoA Dehydrogenase Deficiency (MADD) as a Cause of Rhabdomyolysis. *Mol. Genet. Metab.* **2007**, *91*, 362–369. [CrossRef]
64. Fabius, L.S.; Westermann, C.M. Evidence-Based Therapy for Atypical Myopathy in Horses. *Equine Vet. Educ.* **2018**, *30*, 616–622. [CrossRef]
65. González-Medina, S.; Ireland, J.L.; Piercy, R.J.; Newton, J.R.; Votion, D.M. Equine Atypical Myopathy in the UK: Epidemiological Characteristics of Cases Reported from 2011 to 2015 and Factors Associated with Survival. *Equine Vet. J.* **2017**, *49*, 746–752. [CrossRef] [PubMed]
66. Weinert-Nelson, J.R.; Biddle, A.S.; Williams, C.A. Fecal Microbiome of Horses Transitioning between Warm-Season and Cool-Season Grass Pasture within Integrated Rotational Grazing Systems. *Anim. Microbiome* **2022**, *4*, 41. [CrossRef] [PubMed]
67. Niu, Q.; Pu, G.; Fan, L.; Gao, C.; Lan, T.; Liu, C.; Du, T.; Kim, S.W.; Niu, P.; Zhang, Z.; et al. Identification of Gut Microbiota Affecting Fiber Digestibility in Pigs. *Curr. Issues Mol. Biol.* **2022**, *44*, 4557–4569. [CrossRef] [PubMed]
68. De Vos, W.M.; Tilg, H.; Van Hul, M.; Cani, P.D. Gut Microbiome and Health: Mechanistic Insights. *Gut* **2022**, *71*, 1020–1032. [CrossRef] [PubMed]
69. Hills, R.D.; Pontefract, B.A.; Mishcon, H.R.; Black, C.A.; Sutton, S.C.; Theberge, C.R. Gut Microbiome: Profound Implications for Diet and Disease. *Nutrients* **2019**, *11*, 1613. [CrossRef]
70. Pellegrino, A.; Coppola, G.; Santopaolo, F.; Gasbarrini, A.; Ponziani, F.R. Role of Akkermansia in Human Diseases: From Causation to Therapeutic Properties. *Nutrients* **2023**, *15*, 1815. [CrossRef] [PubMed]
71. Garcia-Mazcorro, J.F.; Minamoto, Y.; Kawas, J.R.; Suchodolski, J.S.; de Vos, W.M. Akkermansia and Microbial Degradation of Mucus in Cats and Dogs: Implications to the Growing Worldwide Epidemic of Pet Obesity. *Vet. Sci.* **2020**, *7*, 44. [CrossRef] [PubMed]
72. Hirayama, M.; Ohno, K. Parkinson's Disease and Gut Microbiota. *Ann. Nutr. Metab.* **2021**, *77*, 28–35. [CrossRef]
73. Costa, M.C.; Weese, J.S. The Equine Intestinal Microbiome. *Anim. Health Res. Rev./Conf. Res. Work. Anim. Dis.* **2012**, *13*, 121–128. [CrossRef]
74. Merritt, A.M.; Julliand, V. Gastrointestinal Physiology. In *Equine Applied and Clinical Nutrition*; Geor, R.J., Harris, P.A., Coenen, M., Eds.; W.B. Saunders: Philadelphia, PA, USA, 2013; pp. 3–32.
75. Feng, P.C.; Patrick, S.J. Studies of the Action of Hypoglycin-A, an Hypoglycaemic Substance. *J. Pharmacol.* **1958**, *13*, 125–130. [CrossRef]
76. Von Holt, C. Methylenecyclopropaneacetic Acid, a Metabolite of Hypoglycin. *Biochim. Biophys. Acta* **1966**, *3*, 1–10. [CrossRef]
77. Ikeda, Y.; Tanaka, K. Selective Inactivation of Various Acyl-CoA Dehydrogenases by (Methylenecyclopropyl)Acetyl-CoA. *Biochim. Biophys. Acta (BBA) Protein Struct. Mol. Enzymol.* **1990**, *1038*, 216–221. [CrossRef]
78. Wexler, H.M. Bacteroides: The Good, the Bad, and the Nitty-Gritty. *Clin. Microbiol. Rev.* **2007**, *20*, 593–621. [CrossRef]
79. Zafar, H.; Saier, M.H. Gut Bacteroides Species in Health and Disease. *Gut Microbes* **2021**, *13*, 1848158. [CrossRef] [PubMed]
80. Tett, A.; Pasolli, E.; Masetti, G.; Ercolini, D.; Segata, N. Prevotella Diversity, Niches and Interactions with the Human Host. *Nat. Rev. Microbiol.* **2021**, *19*, 585–599. [CrossRef] [PubMed]
81. Nie, Q.; Sun, Y.; Li, M.; Zuo, S.; Chen, C.; Lin, Q.; Nie, S. Targeted Modification of Gut Microbiota and Related Metabolites via Dietary Fiber. *Carbohydr. Polym.* **2023**, *316*, 120986. [CrossRef] [PubMed]
82. Massey, L.K.; Sokatch, J.R.; Conrad, R.S. Branched-Chain Amino Acid Catabolism in Bacteria. *Bacteriol. Rev.* **1976**, *40*, 42–54. [CrossRef] [PubMed]
83. Suryawan, A.; Hawes, J.W.; Harris, R.A.; Shimomura, Y.; Jenkins, A.E.; Hutson, S.M. A Molecular Model of Human Branched-Chain Amino Acid Metabolism. *Am. J. Clin. Nutr.* **1998**, *1*, 72–81. [CrossRef]
84. Hutson, S.M.; Hall, T.R. Identification of the Mitochondrial Branched Chain Aminotransferase as a Branched Chain alpha-Keto Acid Transport Protein. *J. Biol. Chem.* **1993**, *268*, 3084–3091. [CrossRef] [PubMed]
85. Hutson, S.M.; Fenstermacher, D.; Mahar, C. Role of Mitochondrial Transamination in Branched Chain Amino Acid Metabolism. *J. Biol. Chem.* **1988**, *263*, 3618–3625. [CrossRef]
86. Hutson, S.M.; Wallinn, R.; Hall, T.R. Identification of Mitochondrial Branched Chain Aminotransferase and Its Isoforms in Rat Tissues. *J. Biol. Chem.* **1992**, *267*, 15681–15686. [CrossRef] [PubMed]

87. Hall, T.R.; Wallinq, R.; Reinhartll, G.D.; Hutson, S.M. Branched Chain Aminotransferase Isoenzymes. Purification and Characterization of the Rat Brain Isoenzyme. *J. Biol. Chem.* **1993**, *268*, 3092–3098. [CrossRef]
88. Brosnan, J.T.; Brosnan, M.E. Branched-Chain Amino Acids: Metabolism, Physiological Function, and Application. *J. Nutr.* **2006**, *136*, 207S–211S. [CrossRef] [PubMed]
89. Neinast, M.; Murashige, D.; Arany, Z. Branched Chain Amino Acids. *Annu. Rev. Physiol.* **2018**, *26*, 139–164. [CrossRef] [PubMed]
90. Kadowaki, H.; Knox, W.E. Cytoslic and Mitochondrial Isoenzymes of Branched-Chain Amino Acid Aminotransferase during Development of the Rat. *Biochem. J.* **1982**, *202*, 777–783. [CrossRef] [PubMed]
91. Hutson, S. Structure and Function of Branched Chain Aminotransferases. *Prog. Nucleic Acid. Res. Mol. Biol.* **2001**, *70*, 175–206. [CrossRef] [PubMed]
92. Chen, C.; Naveed, H.; Chen, K. Research Progress on Branched-Chain Amino Acid Aminotransferases. *Front. Genet.* **2023**, *14*, 1233669. [CrossRef]
93. Bezsudnova, E.Y.; Boyko, K.M.; Popov, V.O. Properties of Bacterial and Archaeal Branched-Chain Amino Acid Aminotransferases. *Biochemistry* **2017**, *82*, 1572–1591. [CrossRef]
94. Harper, A.E.; Miller, R.H.; Block, K.P. Branched-Chain Amino Acid Metabolism. *Ann. Rev. Nutr.* **1984**, *4*, 409–454. [CrossRef] [PubMed]
95. Chuang, D.T. Molecular Studies of Mammalian Branched-Chain a-Keto Acid Dehydrogenase Complexes: Domain Structures, Expression, and Inborn Errors. *Ann. N. Y. Acad. Sci.* **1989**, *573*, 137–154. [CrossRef]
96. Sykes, P.J.; Burns, G.; Menard, J.; Hatter, K.; Sokatch, J.R. Molecular Cloning of Genes Encoding Branched-Chain Keto Acid Dehydrogenase of Pseudomonas Putida. *J. Bacteriol.* **1987**, *169*, 1619–1625. [CrossRef] [PubMed]
97. Perham, R.N.; Lowe, P.N. Isolation and Properties of the Branched-Chain 2-Keto Acid and Pyruvate Dehydrogenase Multienzyme Complex from Bacillus Subtilis. *Methods Enzymol.* **1988**, *166*, 330–342. [PubMed]
98. van der Kolk, J.H.; Wijnberg, I.D.; Westermann, C.M.; Dorland, L.; de Sain-van der Velden, M.G.M.; Kranenburg, L.C.; Duran, M.; Dijkstra, J.A.; van der Lugt, J.J.; Wanders, R.J.A.; et al. Equine Acquired Multiple Acyl-CoA Dehydrogenase Deficiency (MADD) in 14 Horses Associated with Ingestion of Maple Leaves (Acer Pseudoplatanus) Covered with European Tar Spot (Rhytisma Acerinum). *Mol. Genet. Metab.* **2010**, *101*, 289–291. [CrossRef] [PubMed]
99. Sponseller, B.T.; Valberg, S.J.; Schultz, N.E.; Bedford, H.; Wong, D.M.; Kersh, K.; Shelton, G.D. Equine Multiple Acyl-CoA Dehydrogenase Deficiency (MADD) Associated with Seasonal Pasture Myopathy in the Midwestern United States. *J. Vet. Intern. Med.* **2012**, *26*, 1012–1018. [CrossRef]
100. Wood, J.C.; Magera, M.J.; Rinaldo, P.; Reed Seashore, M.; Strauss, A.W.; Friedman, A. Diagnosis of Very Long Chain Acyl-Dehydrogenase Deficiency From an Infant's Newborn Screening Card. *Pediatrics* **2001**, *108*, E19. [CrossRef]
101. Dambrova, M.; Makrecka-Kuka, M.; Kuka, J.; Vilskersts, R.; Nordberg, D.; Attwood, M.M.; Smesny, S.; Sen, Z.D.; Guo, A.C.; Oler, E.; et al. Acylcarnitines: Nomenclature, Biomarkers, Therapeutic Potential, Drug Targets, and Clinical Trials. *Pharmacol. Rev.* **2022**, *74*, 506–551. [CrossRef]
102. Verheyen, T.; Decloedt, A.; de Clercq, D.; van Loon, G. Cardiac Changes in Horses with Atypical Myopathy. *J. Vet. Intern. Med.* **2012**, *26*, 1019–1026. [CrossRef]
103. Tucci, S.; Flögel, U.; Hermann, S.; Sturm, M.; Schäfers, M.; Spiekerkoetter, U. Development and Pathomechanisms of Cardiomyopathy in Very Long-Chain Acyl-CoA Dehydrogenase Deficient (VLCAD-/-) Mice. *Biochim. Biophys. Acta Mol. Basis. Dis.* **2014**, *1842*, 677–685. [CrossRef] [PubMed]
104. Aguer, C.; McCoin, C.S.; Knotts, T.A.; Thrush, A.B.; Ono-Moore, K.; McPherson, R.; Dent, R.; Hwang, D.H.; Adams, S.H.; Harper, M.E. Acylcarnitines: Potential Implications for Skeletal Muscle Insulin Resistance. *FASEB J.* **2015**, *29*, 336–345. [CrossRef]
105. Garber, A.; Hastie, P.; Murray, J.A. Factors Influencing Equine Gut Microbiota: Current Knowledge. *J. Equine Vet. Sci.* **2020**, *88*, 102943. [CrossRef] [PubMed]
106. Votion, D.M.; François, A.C.; Kruse, C.; Renaud, B.; Farinelle, A.; Bouquieaux, M.C.; Marcillaud-pitel, C.; Gustin, P. Answers to the Frequently Asked Questions Regarding Horse Feeding and Management Practices to Reduce the Risk of Atypical Myopathy. *Animals* **2020**, *10*, 365. [CrossRef] [PubMed]
107. Salem, S.E.; Maddox, T.W.; Berg, A.; Antczak, P.; Ketley, J.M.; Williams, N.J.; Archer, D.C. Variation in Faecal Microbiota in a Group of Horses Managed at Pasture over a 12-Month Period. *Sci. Rep.* **2018**, *8*, 8510. [CrossRef] [PubMed]
108. Theelen, M.J.P.; Luiken, R.E.C.; Wagenaar, J.A.; van Oldruitenborgh-Oosterbaan, M.M.S.; Rossen, J.W.A.; Zomer, A.L. The Equine Faecal Microbiota of Healthy Horses and Ponies in The Netherlands: Impact of Host and Environmental Factors. *Animals* **2021**, *11*, 1762. [CrossRef] [PubMed]
109. Fernandes, K.A.; Gee, E.K.; Rogers, C.W.; Kittelmann, S.; Biggs, P.J.; Bermingham, E.N.; Bolwell, C.F.; Thomas, D.G. Seasonal Variation in the Faecal Microbiota of Mature Adult Horses Maintained on Pasture in New Zealand. *Animals* **2021**, *11*, 2300. [CrossRef] [PubMed]
110. Chaucheyras-Durand, F.; Sacy, A.; Karges, K.; Apper, E. Gastro-Intestinal Microbiota in Equines and Its Role in Health and Disease: The Black Box Opens. *Microorganisms* **2022**, *10*, 2517. [CrossRef]

111. de Bustamante, M.M.; Plummer, C.; Macnicol, J.; Gomez, D. Impact of Ambient Temperature Sample Storage on the Equine Fecal Microbiota. *Animals* **2021**, *11*, 819. [CrossRef]
112. Beckers, K.F.; Schulz, C.J.; Childers, G.W. Rapid Regrowth and Detection of Microbial Contaminants in Equine Fecal Microbiome Samples. *PLoS ONE* **2017**, *12*, e0187044. [CrossRef] [PubMed]
113. Costa, M.; Weese, J.S. Methods and Basic Concepts for Microbiota Assessment. *Vet. J.* **2019**, *249*, 10–15. [CrossRef]
114. Dai, Z.-L.; Wu, G.; Zhu, W.Y. Amino Acid Metabolism in Intestinal Bacteria- Links between Gut Ecology and Host Health. *Front. Biosci.* **2011**, *16*, 1768–1786. [CrossRef] [PubMed]
115. Dai, Z.L.; Zhang, J.; Wu, G.; Zhu, W.Y. Utilization of Amino Acids by Bacteria from the Pig Small Intestine. *Amino Acids* **2010**, *39*, 1201–1215. [CrossRef] [PubMed]
116. Leng, J.; Walton, G.; Swann, J.; Darby, A.; La Ragione, R.; Proudmana, C. "Bowel on the Bench": Proof of Concept of a Three-Stage in Vitro Fermentation Model of the Equine Large Intestine. *Appl. Environ. Microbiol.* **2020**, *86*, 1–16. [CrossRef]
117. Molly, K.; Woestyne, M.V.; Smet, I.D.; Verstraete, W. Validation of the Simulator of the Human Intestinal Microbial Ecosystem (SHIME) Reactor Using Microorganism-Associated Activities. *Microb. Ecol. Health Dis.* **1994**, *7*, 191–200. [CrossRef]
118. Venema, K.; Van Den Abbeele, P. Experimental Models of the Gut Microbiome. *Best Pract. Res. Clin. Gastroenterol.* **2013**, *27*, 115–126. [CrossRef] [PubMed]

animals

Article

Use of Hair as Matrix for Trace Elements Biomonitoring in Cattle and Roe Deer Sharing Pastures in Northern Italy

Susanna Draghi [1,*], Nour Elhouda Fehri [1], Fatma Ateş [2], Nural Pastacı Özsobacı [3], Duygu Tarhan [4], Bengü Bilgiç [5], Banu Dokuzeylül [5], Çağla Parkan Yaramış [6], Alev Meltem Ercan [3], Mehmet Erman Or [5], Petra Cagnardi [1], Gabriele Brecchia [1], Giulio Curone [1,*] and Federica Di Cesare [1]

1 Department of Veterinary Medicine and Animal Sciences, University of Milan, 26900 Lodi, Italy; nour.fehri@unimi.it (N.E.F.); petra.cagnardi@unimi.it (P.C.); gabriele.brecchia@unimi.it (G.B.); federica.dicesare@unimi.it (F.D.C.)

2 Department of Biophysics, Faculty of Medicine, Istanbul Beykent University, Istanbul 34398, Turkey; fatmaalkan@beykent.edu.tr

3 Department of Biophysics, Cerrahpasa Faculty of Medicine, Istanbul University-Cerrahpasa, Fatih, Istanbul 34098, Turkey; n.pastaciozsobaci@iuc.edu.tr (N.P.Ö.); meltem@iuc.edu.tr (A.M.E.)

4 Department of Biophysics, School of Medicine, Bahcesehir University, Istanbul 34734, Turkey; duygu.tarhan@bau.edu.tr

5 Department of Internal Medicine, Faculty of Veterinary Medicine, Istanbul University-Cerrahpasa, Istanbul 34320, Turkey; bengu.bilgic@iuc.edu.tr (B.B.); b9eylul@iuc.edu.tr (B.D.); ermanor@iuc.edu.tr (M.E.O.)

6 Department of Plant and Animal Production, Vocational School of Veterinary Medicine, Istanbul University-Cerrahpasa, Istanbul 34320, Turkey; cparkan@iuc.edu.tr

* Correspondence: susanna.draghi@unimi.it (S.D.); giulio.curone@unimi.it (G.C.)

Simple Summary: This study investigates using hair analysis to monitor potentially toxic elements (PTEs) in cattle and roe deer sharing pastures in Northern Italy. PTEs include essential and non-essential elements that, if unbalanced in organisms, can lead to health issues. Hair analysis is a non-invasive method that allows retrospective evaluation of PTE exposure. Aluminum, As, Cd, Cr, Ni, Pb, Cu, Mg, Fe, and Zn were measured. Findings indicate significantly higher As levels in roe deer due to selective feeding, while Cd and Pb levels align with other studies. Cattle have lower Cu, Fe, and Zn levels, likely due to dietary differences. Elevated Cr and Ni in cattle suggest contamination or physiological differences. Hair analysis proves valuable for monitoring environmental PTE exposure, emphasizing interspecies differences and the potential of both animals as bioindicators.

Abstract: Intensive cattle breeding's environmental challenges are prompting shifts to extensive, pasture-based systems, influencing nutrient and pollutant uptake. PTEs are essential and non-essential elements, regularly found in the environment and organisms, and in which unbalances lead to health issues. Hair analysis, a non-invasive method, provides retrospective PTE exposure evaluation. This study aims to understand exposure and species-specific accumulation patterns of PTEs in cattle and roe deer sharing pastures in Northern Italy using the hair analysis. Aluminum, As, Cd, Cr, Ni, Pb, Cu, Mg, Fe, and Zn were quantified through the use of ICP-OES. Findings show As levels significantly higher in roe deer due to their selective feeding, while Cd and Pb levels align with other studies. Essential elements like Cu, Fe, and Zn are lower in cattle, possibly due to diet differences. Higher Cr and Ni levels in cattle suggest contamination or physiological differences in accumulation patterns. In conclusion, hair analysis is valuable for monitoring environmental PTE exposure, highlighting significant interspecies differences and the potential of both animals as bioindicators in shared grazing areas.

Keywords: bioindicators; eco-toxicology; environmental toxicology; PTEs; wildlife

1. Introduction

The traditional semi-extensive farming system of cattle in Northern Italy provides the use of pastures from spring to autumn, during which the animals feed on fresh forage, and indoor housing during winter with a diet based on hay obtained from lands adjacent to pastures and farms [1]. Inevitably, concerning the nutrients and the pollutants acquired through the diet, this system creates a strong dependence on the territory in which the animals are raised [2]. Among the substances to consider in this context, particular importance is given to the essential and non-essential elements, also called potentially toxic elements (PTEs). The term PTEs is attributed to the fact that many of them are essential for carrying out physiological activities but become toxic at high concentrations; others are non-essential or toxic, but even in this case, their toxicity depends on the concentration. Due to their importance, over the decades PTEs have been extensively studied in both humans and animals, as their imbalances result in serious health issues [3].

The dual nature of PTEs makes it necessary to monitor their presence and quantities. Given that the absorption of any substance by a living organism depends on various intrinsic factors, both related to the animal (e.g., species, gender, or age) and the substance (chemical–physical properties), monitoring their presence and quantity in soil, water, feed, or food is insufficient to define real exposure [4]. That is why, for many years, environmental toxicology studies have employed biomonitoring. The term biomonitoring is defined as "the measurement and assessment of toxicants or their metabolites either in tissues, secreta, excreta, expired air, or any combination of these to evaluate exposure and health risk". Since decades, wild animals have been used for this type of study [5] as considered effective tools due to their direct exposure to the natural environment [6,7]. Many wild mammals have been identified as ideal bioindicators, among which the roe deer (Capreolus capreolus) stands out [8,9]. Being an herbivorous ruminant with limited home ranges, it can effectively indicate the presence of environmental contaminants in a specific territory [10]. Considering the reasons behind its selection as an ideal bioindicator, the potential for using pasture-raised cattle as biomonitoring tools also becomes obvious. Pasture-raised cattle, with home ranges comparable to those of roe deer (16–20 hectares) [11], share a similar characteristic, i.e., they are ruminants whose diet relies on local forage. Despite differences in body size, life expectancy, and human intervention in diet, particularly feed supplementation in the case of cattle, both species could be able to offer valuable insights into PTE accumulation due to intrinsic species-specific factors when coexisting in grazing areas.

An effective method for assessing PTE levels in animals involves the examination of hair [12]. Hair, once separated from the epidermis, serves as a metabolically inert biological matrix that is chemically uniform [13]. Its non-invasive collection from both wild and domestic animals is straightforward, and its growth pattern permits a retrospective analysis of the examined element for several months post-collection [14]. Moreover, the sampling can be repeated on the same skin area, allowing for repeated measurements [2]. The hair analysis facilitates the determination of prolonged exposure to both trace elements and heavy metals, owing to the chelation capability of the sulfhydryl group (–SH) in cysteine [14].

Therefore, the purpose of this study is to employ the hair of both cattle and roe deer from the same area for biomonitoring the quantity of some PTEs, i.e., Al, As, Cd, Cr, Cu, Fe, Mg, Ni, Pb, and Zn. Secondly, it aims to compare the differences in PTE accumulation between the two species to understand the influence of the species and the rearing environment. Finally, it would determine whether roe deer exhibit accumulation patterns similar to cattle and can be used as a biomonitoring tool to assess the health of an area potentially designated for grazing.

2. Materials and Methods

2.1. Animal and Hair Collection

Before starting this study, the approval of the Institutional Animal Care and Use Committee of Università degli Studi di Milano (Permission OPBA_26_2022) as a non-experimental project was requested and obtained.

At first, a specific area of Northern Italy with the presence of semi-extensive cattle breeding farms and a well-structured hunting activity was chosen (Figure 1). From a farm with pastures distributed throughout the sampling area, 40 healthy cows were randomly selected. All the cows were females, multiparous (average 2.2 deliveries), aged averagely 4.7 years; all animals were of dual-purpose, producing both milk and meat, raised on pasture from April to November and kept in barns from December to March. During the months they were indoors, the cows were fed with hay obtained from areas near the pastures. Throughout the year, each cow was fed a complete feed in the amount of 2.5 kg per day (the specific characteristics of the feed, i.e., ingredients, vitamins, trace elements, and analytical data, are given in Tables S1–S3). To standardize, the 40 sampled roe deer were all females with an average age of 3 years, randomly selected from the hunting plan. The age of the animals was estimated through the evaluation of dental eruptions and erosion.

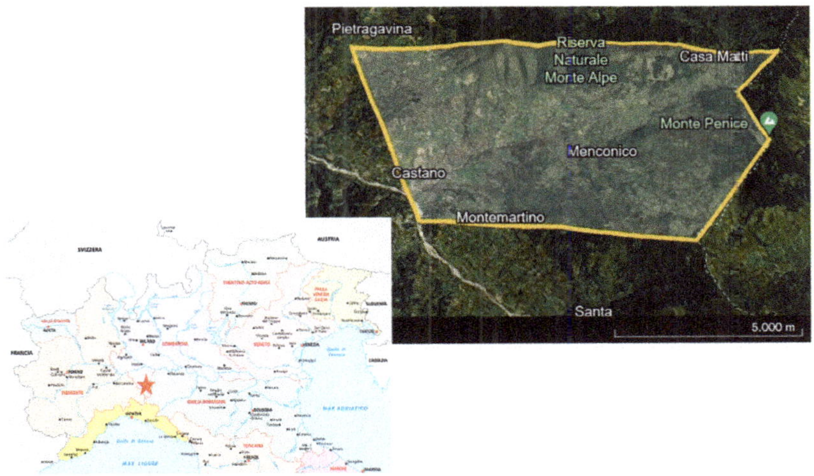

Figure 1. Map of the sampling area. The red star represents the location of the sampling area on the map of northern Italy.

In both species, the hair samples were collected from an area of 10 cm^2 on the left side of the animal at the level of the costal arch. For hair collection, an electric shaver was used, and the hairs were cut at the base. The sampling procedure was conducted separately for the two species. The cows' hair was sampled directly in the pasture over multiple days to cover the entire hunting period during which roe deer hair was collected (June–October). The roe deer hair was sampled during regular hunting activities at the game meat processing center. After sampling, the hair was placed in specific plastic bags and stored at room temperature, in a dry environment, and protected from sources of light.

2.2. Analysis of PTEs

Aluminum (Al), arsenic (As), cadmium (Cd), chromium (Cr), nickel (Ni), lead (Pb), copper (Cu), magnesium (Mg), iron (Fe), and zinc (Zn) elements were carried out by using an inductively coupled plasma-optical emission spectrophotometer (ICP-OES; Thermo iCAP 6000series) at Istanbul University-Cerrahpasa. ICP-OES device parameters for determining PTEs are presented in Table 1. The quality assurance of ICP-OES analysis was ensured through the utilization of appropriate test solutions, each containing 2000 ppm (mg/L) for every element under examination, sourced from Chem-Lab NV. Standard solutions for all PTEs were meticulously prepared by diluting standards with 1000 ppm (mg/L) concentrations for each element, also obtained from Chem-Lab NV, in deionized water. A 3-point calibration was conducted using standard and blank solutions as reference materials. This process resulted in the acquisition of reproducible and linear calibration curves for analysis, with a determination of the correlation coefficient for each measured element. The calibration graph was generated using blank and standard solutions on the ICP-OES device, facilitating the subsequent elemental analysis of the prepared samples. To ensure precision, each measurement was replicated three times, and the results were averaged for comprehensive analysis. This rigorous methodology is aimed at maintaining the integrity and accuracy of the ICP-OES analytical process. In the study, the appropriate wavelengths (Table 2) of all elements were used for the analysis by the ICP-OES device. According to methodology, before starting the analyses, the hair was washed with acetone, three times with deionized water, and finally with acetone to remove surface impurities and any adhesive contamination present on the surface of the hair. Then, all samples underwent dissolution in a drying oven (Heraeus W.C., Hanau, Germany) at 180 °C. The average sample weight used was 0.143 g. This process involved the addition of 2 mL of 65% nitric acid (Merck, Darmstadt, Germany) and 1 mL of 60% perchloric acid (Panreac, Barcelona, Spain). After cooling to room temperature, the suspension was vortexed, and distilled water was added to the samples, reaching a total volume of 10 mL. The quantification of the concentration of each element was conducted individually, with consideration given to the weight of the respective sample. This systematic approach ensured a precise and methodical assessment of the elemental concentrations in the samples. The recovery of the analyzed quality control was between 96% and 108%. Table 3 shows the results of the ICP-OES method validation. Trace and toxic element levels were expressed as $mg \cdot kg^{-1}$ of sample wet weight.

Table 1. ICP-OES device parameters for determination of elements.

Parameters	Assigned Value
Plasma gas flow rate	15 L/min
Argon carrier flow rate	0.5 L/min
Sample flow rate	1.51 L/min
The speed of peristaltic pump	100 rpm
RF Power	1150 W

Table 2. Wavelengths used in the analysis for each element.

Elements	Wavelength (nm)
Aluminum (Al)	167.070
Arsenic (As)	189.042
Cadmium (Cd)	228.802
Chromium (Cr)	267.716
Nickel (Ni)	341.476
Lead (Pb)	220.353
Copper (Cu)	324.754
Magnesium (Mg)	285.213
Iron (Fe)	259.940
Zinc (Zn)	206.200

Table 3. Results of ICP-OES method validation for elements.

Elements	Quality Control (QC)	LOD	LOQ	Expected Concentration	Measured Concentration (n = 3) (ppm)	Precision (RSD%)	Recovery (%)
Al	QC-1 QC-2	0.001	0.003	2.500 5.000	2.420 5.000	0.168 0.144	96.8 100
As	QC-1 QC-2 QC-3 QC-4	0.000	0.001	0.050 0.100 0.500 1.000	0.050 0.100 0.490 0.960	5.684 0.974 4.248 1.274	100 100 98 96
Cd	QC-1 QC-2	0.008	0.014	0.500 1.000	0.536 0.963	0.752 0.213	107.2 96.3
Cr	QC-1 QC-2	0.003	0.008	0.050 0.100	0.052 0.097	0.726 0.816	104 97
Ni	QC-1 QC-2 QC-3 QC- 4	0.000	0.002	0.050 0.100 0.500 1.000	0.052 0.099 0.490 0.990	6.844 2.474 0.838 0.124	104 99 98 99
Pd	QC-1 QC-2	0.000	0.004	0.500 1.000	0.490 1.010	3.332 0.451	98 101
Cu	QC-1 QC-2	0.002	0.006	0.500 1.000	0.509 0.990	0.32 0.675	101.8 99
Mg	QC-1 QC-2	0.000	0.001	0.500 1.000	0.524 0.973	1.895 0.539	104.8 97.3
Fe	QC-1 QC-2	0.003	0.004	0.500 1.000	0.490 0.990	5.772 0.846	98 99
Zn	QC-1 QC-2	0.003	0.009	0.500 1.000	0.540 1.000	0.756 0.343	108 100

QC: quality control; LOD: limit of detection; LOQ: limit of quantitation; RSD: relative standard deviation.

2.3. Statistical Analysis

Statistical analyses were conducted using GraphPad InStat 8 software (version 8.0.2). Initially, the data were categorized based on species (bovine = 40 and roe deer = 40). Subsequently, the Shapiro–Wilk normality test was performed. As the data were found to be non-normally distributed, a non-parametric statistical test (Mann–Whitney test) was applied for comparisons between the two categories.

3. Results

During this study, statistically significant differences were highlighted between the two species in the concentration of PTEs in the hair. In cattle, compared to roe deer hair, the content of Cr, Ni, Mg, and Zn was found to be significantly higher with a p-value < 0.001. In the case of Al, As, Pb, and Cu, significantly higher concentrations were recorded in roe deer hair with p < 0.001, except for Pb, which showed p = 0.051. The average median with percentiles, minimum, and maximum data are reported in Table 4. Graphical representations of comparisons between the two species and statistical significances are shown in Figure 2.

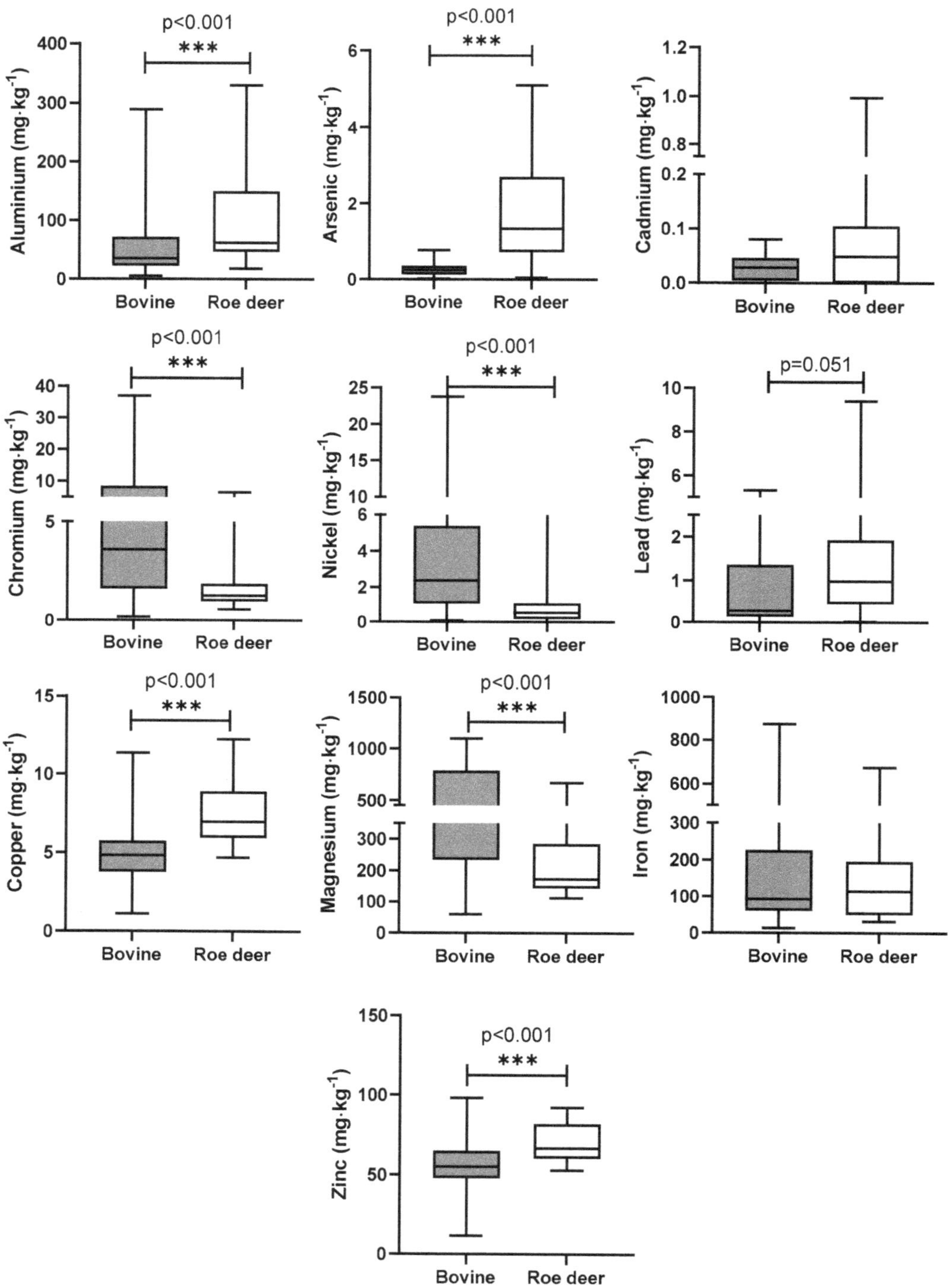

Figure 2. Graphical representation comparing the concentrations of PTEs in the hair of both cattle and roe deer. ***: extremely significant.

Table 4. Mean, median, minimum, and maximum concentrations of PTEs quantified in bovine and roe deer hair. The values are expressed in mg·kg^{-1}.

Element	Species	Mean ± SD	Min–Max	Percentile			*p* Value
				25th	Median	75th	
Al	Bovine	61.34 ± 71.86	5.26–289.03	22.74	35.00	70.12	<0.001
	Roe deer	98.64 ± 77.03	17.88–329.6480	46.93	61.88	144.04	
As	Bovine	0.26 ± 0.17	0.01–0.77	0.14	0.24	0.34	<0.001
	Roe deer	1.84 ± 1.51	0.05–5.1	0.78	1.34	2.58	
Cd	Bovine	0.03 ± 0.03	0–0.08	0.01	0.03	0.04	0.167
	Roe deer	0.08 ± 0.16	0–0.99	0.00	0.05	0.10	
Cr	Bovine	5.66 ± 6.68	0.19–36.91	1.75	3.62	8.25	<0.001
	Roe deer	1.61 ± 1.04	0.59–6.58	1.04	1.31	1.89	
Ni	Bovine	3.73 ± 4.34	0.09–23.78	1.38	2.39	5.33	<0.001
	Roe deer	0.83 ± 1.13	0–6.78	0.19	0.54	1.05	
Pb	Bovine	0.97 ± 1.39	0–5.32	0.16	0.28	1.30	0.051
	Roe deer	1.39 ± 1.63	0.01–9.42	0.46	0.99	1.92	
Cu	Bovine	4.92 ± 2.2	1.11–11.35	3.84	4.83	5.60	<0.001
	Roe deer	7.49 ± 2.09	4.71–12.21	5.95	6.95	8.84	
Mg	Bovine	490.93 ± 306.01	59.96–1097.08	258.56	384.77	777.63	<0.001
	Roe deer	226.89 ± 120.14	112.24–675.22	147.23	173.84	279.87	
Fe	Bovine	175.13 ± 196.38	13.09–873.56	65.24	93.32	212.01	0.984
	Roe deer	156.11 ± 143.39	0–674.14	50.21	113.77	188.79	
Zn	Bovine	54.43 ± 19.34	11.71–98.12	48.11	55.06	64.90	<0.001
	Roe deer	69.34 ± 11.54	52.78–92.03	60.66	66.79	78.96	

4. Discussion

Worldwide, hair analysis is considered a suitable method for the assessment of the health status and the mineral metabolism of animals. In this study, aluminum (Al), arsenic (As), cadmium (Cd), chromium (Cr), nickel (Ni), lead (Pb), copper (Cu), magnesium (Mg), iron (Fe), and zinc (Zn) were quantified in the hair of both cattle and roe deer grazing within the same geographical area. The aim was to use them as bioindicators for the presence and environmental quantities of PTEs and subsequently compare the concentrations between the two species to identify any interspecies differences. This approach is useful in understanding the potential physiological, ethological, and dietary influences on PTE accumulation in hair [2]. The cattle were grazed in pastures from March to November, thus sharing the home range with the roe deer hunted in that area. The use of hair to assess mineral metabolism and the health status of animals has been employed in various domestic species, such as dogs [15], cats [16], horses [17], and cattle [18], as well as in various wild species like bison [19], red deer [20], mule deer [21], and even roe deer [12]. Technically, interpreting data regarding PTE content in hair is only possible after comparison with species-specific reference values [22]. In our case, reference values exist only for cattle [23,24].

Comparing the median values identified in the hair of cattle enrolled in this study with the published reference values [23,24], both similarities and differences were revealed. The concentrations of As, Cd, and Pb turned out to be very similar compared with the reference values reported by Miroshnikov et al. [23] in Hereford cows. In the hair of roe deer, a similar study by Cygan-Szczegielniak et al. [12] conducted in Poland showed a higher concentration of Cd and a lower concentration of Pb compared to the concentrations identified in the roe deer hair in our study. In the study of Cygan-Szczegielniak et al.,

As has not been evaluated in roe deer hair. Comparing the concentration of As, Cd, and Pb between the two species, only As is significantly higher in roe deer hair ($p < 0.001$; Figure 2). As is a metalloid, both naturally present in the environment and derived from human activities. It is regularly traced in surface and groundwater, and its quantity is due to the mineral–water interaction [25]. Considering the natural presence of this element in rivers and streams [26], the identified difference between the two species could be due to their behavior and rearing conditions. Indeed, deer have larger home ranges and a larger number of watering points. In contrast, cattle, confined to fenced pastures with restricted access to watering points, may experience lower exposure to arsenic if the water sources have lower concentrations, hence resulting in reduced arsenic accumulation in their hair. Moreover, due to its selective browsing habits [27], the roe deer tends to target plant buds, roots, and seeds, which are arsenic-accumulating organs, thus elevating its susceptibility to arsenic uptake [28].

Regarding Cu, Fe, and Zn in cattle, the concentrations obtained in our study were approximately half compared to those obtained in the study by Miroshnikov et al. [24] on Holstein cows. These concentration differences between our and the reference values, particularly in elements considered essential for cattle, could be attributed to variations in the biogeochemical composition of pastures and dietary supplements [29]. Indeed, the cattle used in our study were dual-purpose (milk and meat) that grazed in pastures for 8 months, whereas the cattle used as reference values are Holstein cows bred for milk production and always kept in barns. Regarding the content of Cu, Zn, and Fe in roe deer hair, it was found that the concentration of Cu and Zn identified in our study was much lower compared to that identified by Cygan-Szczegielniak et al. [23], while the Fe content resulted 10 times higher compared to the study of Długaszek et al. [30]. Comparing the content of Cu and Zn between the considered species, they were at significantly higher concentrations in roe deer species. This result is rather controversial, as cattle were administered supplementary feed containing Cu and Zn, and thus higher concentrations were expected at the fur level of this species. However, one possible explanation could be attributed to the different diets. Temporarily setting aside the supplements, cattle primarily feed on fresh or dried forage, while roe deer, as reported in the literature, feed on roots, buds, and leaves of plants and shrubs [31]. In the case of Cu, a study by Yruela et al. [32] reports that the genes responsible for copper uptake, distribution, and storage in plants are expressed at the level of roots, buds, and flowers. The same principle applies to Zn, which tends to be higher in reproductive vegetative tissues as well as in shoots and roots [33].

The concentrations of Cr and Ni found in our cattle were many times higher compared to those identified in the hair of both Hereford cattle raised in pastures [23] and Holstein cows raised in barns [24]. The proposed reference interval of Cr for Hereford cattle raised in pastures is 0.053–1.121 $mg \cdot kg^{-1}$. In the case of Holstein Friesian cattle raised in barns, the interval narrows to 0.0671–0.409 $mg \cdot kg^{-1}$. In our study, the median value of Cr in bovine was 3.620 $mg \cdot kg^{-1}$. Chromium is considered an essential element for ruminants as it plays a role in carbohydrate and lipid metabolism and has beneficial effects on the immune system [34]. However, at excessive concentrations, it has a negative impact on the respiratory and gastrointestinal systems and also exhibits carcinogenic and embryotoxic effects [35]. Chromium is an element that is both naturally present in the Earth's crust and originates from industrial processes [36], and its bioavailability depends on soil composition [37]. Therefore, cattle raised in different areas, grazing on different soils, or fed with forage grown in soils with varying compositions could necessarily have different tissue concentrations of Cr [18]. The concentration of Cr in roe deer hair was similar to studies conducted on the same species and other wild species [38].

The median Ni concentration identified in the hair in our study was 2.391 $mg \cdot kg^{-1}$, whereas the reference intervals were 0.108–0.595 $mg \cdot kg^{-1}$ for lactating cows and 0.280–1.575 $mg \cdot kg^{-1}$ in beef cattle [23,24]. Nickel is not considered an essential element, and its excess is carcinogenic and teratogenic [39]. Similarly for Cr, in the case of Ni, the concentration in roe deer hair was found to be consistent with the concentrations identified in other studies on the same species and on other wild species [38]. It is naturally present in the Earth's crust and, due to its chemical and physical properties, finds widespread use in various applications, including modern metallurgies such as alloy production, electroplating, and the manufacturing of nickel-cadmium batteries [40]. The extensive use of products containing Ni inevitably contributes to environmental pollution [41] and determines the difference in exposure.

Comparing the concentration of Cr and Ni in the hair of cattle and roe deer, it was found that cattle showed a significantly higher concentration of both elements. The sources of exposure are mainly dietary; however, the reason for this difference is unclear. It could be hypothesized contamination of cattle drinking water or supplementary feed, excluding hay, as it is collected in the same area where roe deer live. Alternatively, a physiological difference between the species such as body mass, rumen size, and feeding behavior (indeed, based on feeding behavior, cattle are grazers, while roe deer are selective browsers) could have led cattle to a higher intake of Ni and Cr. Future studies identifying the concentration of Cr and Ni in the feed of both species might help explain this difference.

Aluminum and Mg were not included in the reference values [23,24] but were identified in another study of Linhares et al. [18], where, in the hair of pasture-raised cattle, higher concentrations were found compared to our study. Our lower concentrations of Al and Mg could be probably due to the different soil composition; indeed, cattle enrolled in the study of Linhares et al. belonged to areas with volcanic soils [18].

The Al content identified in the hair of roe deer in this study was found to be much lower compared to that identified in the hair of other wild animals [38]; the difference could be due to intraspecific and interspecific differences, sampling period, and sampling area. For the concentration of Mg, it resulted similar to another study of our research group in roe deer hair [10]. Moreover, there are also studies reporting the concentration of Mg in the fallow deer antlers [42] or in the cranial bone and antlers of red deer [43], but considering the physiological inter-specific differences and the differences between the matrices, the comparison is not possible.

Meanwhile, comparing the concentration of Al in hair between cattle and roe deer, the hair of deer was found to have a significantly higher level. For this element as well, the main source of exposure appears to be the diet. Aluminum primarily accumulates in plant roots [44], so the feeding behavior of deer could explain the higher concentration in their fur compared to cattle. On the other hand, Mg was found to be higher in the hair of cattle, despite the supplementary feed given to them daily not containing it. Therefore, it is possible to hypothesize that this result depends on the physiology of the two species or their body mass.

5. Conclusions

In conclusion, this study highlights the significance of hair analysis as a valuable method for assessing the health status and mineral metabolism of animals, particularly in understanding the accumulation of PTEs. By comparing the concentrations of various elements in the hair of cattle and roe deer, significant interspecies differences were identified, shedding light on potential environmental influences, dietary habits, and physiological factors impacting element accumulation. Further research into the dietary sources of these elements and physiological differences between species is warranted to fully elucidate these findings.

Supplementary Materials: The following supporting information can be downloaded at: https://www.mdpi.com/article/10.3390/ani14152209/s1. Table S1: Composition of the complementary feed administered to bovine. The feed was administered in the quantity of 2.5 kg/day. The ingredient quantity is expressed in kg; Table S2. Composition of the supplement; Table S3. Diet's analytical composition.

Author Contributions: Conceptualization, S.D. and G.C.; methodology, F.A., N.P.Ö., D.T., B.B., B.D., Ç.P.Y. and A.M.E.; software, S.D.; validation, F.A., N.P.Ö., D.T., B.B., B.D., Ç.P.Y., A.M.E. and P.C.; formal analysis, S.D.; investigation, S.D.; resources, F.A., N.P.Ö., D.T. and B.B.; data curation, S.D.; writing—original draft preparation, S.D., P.C. and F.D.C.; writing—review and editing, S.D., P.C. and F.D.C.; visualization, N.E.F.; supervision, S.D., M.E.O., G.B., F.D.C. and G.C.; project administration, G.C., S.D., F.D.C. and M.E.O.; funding acquisition, G.B. and M.E.O. All authors have read and agreed to the published version of the manuscript.

Funding: This research received no external funding.

Institutional Review Board Statement: The animal study protocol was approved by the Ethics Committee of Università degli Studi di Milano (Permission OPBA_26_2022).

Informed Consent Statement: Not applicable.

Data Availability Statement: The data presented in this study are available in the article and Supplementary Materials. Further information is available upon request from the corresponding author.**Aknowledgements:** The authors acknowledge support from the University of Milan through the APC initiative.

Conflicts of Interest: The authors declare no conflicts of interest.

References

1. Battaglini, L.; Bovolenta, S.; Gusmeroli, F.; Salvador, S.; Sturaro, E. Environmental Sustainability of Alpine Livestock Farms. *Ital J. Anim. Sci.* **2014**, *13*, 3155. [CrossRef]
2. Agradi, S.; Munga, A.; Barbato, O.; Palme, R.; Dokuzeylül, B.; Ercan, A.M.; Or, M.E.; Brecchia, G.; Curone, G.; Draghi, S. Goat Hair as a Bioindicator of Environmental Contaminants and Adrenal Activation During Vertical Transhumance. *Front. Vet Sci.* **2023**, *10*, 1274081. [CrossRef] [PubMed]
3. Famuyiwa, A.; Ande, S.; Adeboye, A.; Kehinde, O.; Sulaiman, N.; Otaniyi, O.; Taiwo, A. Pollution characterisation, ecological and human health risks assessment of potentially toxic elements in soils of Ikorodu industrial area of Lagos, Southwestern, Nigeria. *Trans. R. Soc. South Afr.* **2021**, *76*, 189–199. [CrossRef]
4. Draghi, S.; Curone, G.; Pavlovic, R.; Di Cesare, F.; Cagnardi, P.; Fornesi Silva, C.; Pellegrini, A.; Riva, F.; Arioli, F.; Fidani, M. Influence of Area, Age and Sex on Per- and Polyfluorinated Alkyl Substances Detected in Roe Deer Muscle and Liver from Selected Areas of Northern Italy. *Animals* **2024**, *14*, 529. [CrossRef] [PubMed]
5. Dhananjayan, V.; Jayanthi, P.; Ravichandran, B.; Jayakumar, R. *Chapter 21—Biomonitoring and Biomarkers of Pesticide Exposure and Human Health Risk Assessment*; Elsevier Inc.: Amsterdam, The Netherlands, 2022; pp. 563–584.
6. Sobota, S.; Baranowska-Bosiacka, I.; Gutowska, I.; Kupiec, M.; Dusza, E.; Machoy, Z.; Chlubek, D. Biomonitoring of lead and fluoride contamination in forests using chemical analysis of hard tissues of roe deer (*Capreolus capreolus* L.). *Pol. J. Environ. Stud.* **2011**, *20*, 435–443.
7. Draghi, S.; Spinelli, M.; Fontanarosa, C.; Curone, G.; Amoresano, A.; Pignoli, E.; Cagnardi, P.; Vigo, D.; Arioli, F.; Materazzi, S.; et al. Evaluation of the Difference in the Content of Essential and Non-Essential Elements in Wild Boar and Swine Tissues Sampled in the Same Area of Northern Italy. *Animals* **2024**, *14*, 827. [CrossRef] [PubMed]
8. Pavlovic, R.; Draghi, S.; Pellegrini, A.; Fornesi Silva, C.; Di Cesare, F.; Curone, G.; Arioli, F.; Fidani, M. High-Resolution Mass Spectrometry Non-Targeted Detection of Per- and Polyfluoroalkyl Substances in Roe Deer (*Capreolus capreolus*). *Molecules* **2024**, *29*, 617. [CrossRef]
9. Jelenko, I.; Pokorny, B. Historical biomonitoring of fluoride pollution by determining fluoride contents in roe deer (*Capreolus capreolus* L.) antlers and mandibles in the vicinity of the largest Slovene thermal power plant. *Sci. Total Environ.* **2010**, *409*, 430–438. [CrossRef]
10. Draghi, S.; Agradi, S.; Riva, F.; Tarhan, D.; Bilgiç, B.; Dokuzeylül, B.; Ercan, A.M.; Or, M.E.; Brecchia, G.; Vigo, D.; et al. Roe Deer (*Capreolus capreolus*) Hair as a Bioindicator for the Environmental Presence of Toxic and Trace Elements. *Toxics* **2023**, *11*, 49. [CrossRef]
11. Pandini, W.; Cesaris, C. Home range and habitat use of roe deer (*Capreolus capreolus*) reared in captivity and released in the wild. *Hystrix* **1997**, *9*, 45–50. [CrossRef]
12. Cygan-Szczegielniak, D.; Stasiak, K. Effects of age and sex on the content of heavy metals in the hair, liver and the longissimus lumborum muscle of roe deer *Capreolus capreolus* L. *Environ. Sci. Pollut. Res. Int.* **2022**, *29*, 10782–10790. [CrossRef]
13. Cygan-Szczegielniak, D.; Stanek, M.; Stasiak, K.; Roślewska, A.; Janicki, B. The Content of Mineral Elements and Heavy Metals in the Hair of Red Deer (*Cervus elaphus* L.) from Selected Regions of Poland. *Folia Biol.* **2018**, *66*, 133–142. [CrossRef]
14. Perillo, L.; Arfuso, F.; Piccione, G.; Dara, S.; Tropia, E.; Cascone, G.; Licitra, F.; Monteverde, V. Quantification of Some Heavy Metals in Hair of Dairy Cows Housed in Different Areas from Sicily as a Bioindicator of Environmental Exposure—A Preliminary Study. *Animals* **2021**, *11*, 2268. [CrossRef]

15. So, K.-M.; Yoonseok, L.; Jin Duck, B.; Eun Bae, K.; Myung Il, C. Analysis of Ionomic Profiles of Canine Hairs Exposed to Lipopolysaccharide (LPS)-Induced Stress. *Biol. Trace Elem. Res.* **2016**, *172*, 364–371. [CrossRef]
16. Rzymski, P.; Niedzielski, P.; Dąbrowski, P. Assessment of iron in uterine and testicular tissues and hair of free-ranging and household cats. *Pol. J. Vet. Sci.* **2015**, *18*, 677–682. [CrossRef]
17. Ghorbani, A.; Mohit, A.; Darmani Kuhi, H. Effects of Dietary Mineral Intake on Hair and Serum Mineral Contents of Horses. *J. Equine Vet. Sci.* **2015**, *35*, 295–300. [CrossRef]
18. Linhares, D.; Garcia, P.; Pimentel, A.; Gaspar, D.; Rodrigues, A.D.S. Using cattle hair to assess exposure to essential trace elements in volcanic soils. *J. Trace Elem. Med. Biol.* **2024**, *81*, 127345. [CrossRef]
19. Kośla, T.; Skibniewska, E.; Skibniewski, M. The state of bioelements in the hair of free-ranging European bisons from Bialowieza Primeval Forest. *Pol. J. Vet. Sci.* **2011**, *14*, 81. [CrossRef]
20. Oropesa, A.-L.; Ramos, A.; Gómez, L.-J. Toxic and essential metal levels in the hair of red deer (*Cervus elaphus*) and wild boar (Sus scrofa) for monitoring the contamination in protected areas of South-Western Spain. *Environ. Sci. Pollut. Res. Int.* **2022**, *29*, 27430–27442. [CrossRef]
21. Roug, A.; Swift, P.K.; Gerstenberg, G.; Woods, L.W.; Kreuder-Johnson, C.; Torres, S.G.; Puschner, B. Comparison of trace mineral concentrations in tail hair, body hair, blood, and liver of mule deer (*Odocoileus hemionus*) in California. *J. Vet. Diagn. Investig.* **2015**, *27*, 295–305. [CrossRef]
22. Pieper, L.; Schmidt, F.; Müller, A.-E.; Staufenbiel, R. Zinc concentrations in different sample media from dairy cows and establishment of reference values. *Tierarztl. Praxis. Ausg. G Grosstiere/Nutztiere* **2017**, *45*, 213–218. [CrossRef]
23. Miroshnikov, S.A.; Zavyalov, O.A.; Frolov, A.N.; Bolodurina, I.P.; Kalashnikov, V.V.; Grabeklis, A.R.; Tinkov, A.A.; Skalny, A.V. The Reference Intervals of Hair Trace Element Content in Hereford Cows and Heifers (*Bos taurus*). *Biol. Trace Elem. Res.* **2017**, *180*, 56–62. [CrossRef] [PubMed]
24. Miroshnikov, S.A.; Skalny, A.V.; Zavyalov, O.A.; Frolov, A.N.; Grabeklis, A.R. The Reference Values of Hair Content of Trace Elements in Dairy Cows of Holstein Breed. *Biol. Trace Elem. Res.* **2020**, *194*, 145–151. [CrossRef] [PubMed]
25. Panagiotaras, D.; Nikolopoulos, D. Arsenic Occurrence and Fate in the Environment; A Geochemical Perspective. *J. Earth Sci. Clim. Chang.* **2015**, *6*, 1. [CrossRef]
26. Smedley, P.L.; Kinniburgh, D.G. A review of the source, behaviour and distribution of arsenic in natural waters. *Appl. Geochem.* **2002**, *17*, 517–568. [CrossRef]
27. Hewison, A.J.M.; Vincent, J.P.; Angibault, J.M.; Delorme, D.; Laere, G.V.; Gaillard, J.M. Tests of estimation of age from tooth wear on roe deer of known age: Variation within and among populations. *Can. J. Zool.* **1999**, *77*, 58–67. [CrossRef]
28. Bhattacharya, S.; Sharma, P.; Mitra, S.; Mallick, I.; Ghosh, A. Arsenic uptake and bioaccumulation in plants: A review on remediation and socio-economic perspective in Southeast Asia. *Environ. Nanotechnol. Monit. Manag.* **2021**, *15*, 100430. [CrossRef]
29. Rosendahl, S.; Anturaniemi, J.; Vuori, K.A.; Moore, R.; Hemida, M.; Hielm-Björkman, A. Diet and dog characteristics affect major and trace elements in hair and blood of healthy dogs. *Vet. Res. Commun.* **2022**, *46*, 261–275. [CrossRef]
30. Dugaszek, M.; Kopczyski, K. Correlations Between Elements in the Fur of Wild Animals. *Bull. Environ. Contam. Toxicol.* **2014**, *93*, 25–30. [CrossRef]
31. Mussa, P.P.; Aceto, P.; Abba, C.; Sterpone, L.; Meineri, G. Preliminary study on the feeding habits of roe deer (*Capreolus capreolus*) in the western Alps. *J. Anim. Physiol. Anim. Nutr.* **2003**, *87*, 105–108. [CrossRef]
32. Yruela, I. Copper in plants: Acquisition, transport and interactions. *Funct. Plant Biol. FPB* **2009**, *36*, 409–430. [CrossRef] [PubMed]
33. Longnecker, N.E.; Robson, A.D. Distribution and Transport of Zinc in Plants. In *Zinc in Soils and Plants*; Developments in Plant and Soil Sciences; Springer: Dordrecht, The Netherlands, 1993; pp. 79–91.
34. Dworzański, W.; Sembratowicz, I.; Cholewińska, E.; Tutaj, K.; Fotschki, B.; Juśkiewicz, J.; Ognik, K. Effects of Different Chromium Compounds on Hematology and Inflammatory Cytokines in Rats Fed High-Fat Diet. *Front. Immunol.* **2021**, *12*, 614000. [CrossRef] [PubMed]
35. Żarczyńska, K.; Krzebietke, S. The effect of chromium on ruminant health. *J. Elem.* **2020**, *25*, 893–903. [CrossRef]
36. Prasad, S.; Yadav, K.K.; Kumar, S.; Gupta, N.; Cabral-Pinto, M.M.S.; Rezania, S.; Radwan, N.; Alam, J. Chromium contamination and effect on environmental health and its remediation: A sustainable approaches. *J. Environ. Manag.* **2021**, *285*, 112174. [CrossRef] [PubMed]
37. Vanceuren, A.; Pereira, B.; Kaba, A.J.; Titeux, H.; Delmelle, P. Environmental bioavailability of arsenic, nickel and chromium in soils impacted by high geogenic and anthropogenic background contents. *Sci. Total Environ.* **2023**, *902*, 166073. [CrossRef] [PubMed]
38. Squadrone, S.; Robetto, S.; Orusa, R.; Griglione, A.; Falsetti, S.; Paola, B.; Abete, M.C. Wildlife hair as bioindicators of metal exposure. *Biol. Trace Elem. Res.* **2022**, *200*, 5073–5080. [CrossRef] [PubMed]
39. Sigel, A.; Sigel, H.; Sigel, R.K.O. Nickel and Human Health. In *Metal Ions in Life Sciences*; Springer: Amsterdam, The Netherlands, 2013; Volume 13, pp. 321–357.
40. Das, K.K.; Reddy, R.C.; Bagoji, I.B.; Das, S.; Bagali, S.; Mullur, L.; Khodnapur, J.P.; Biradar, M.S. Primary concept of nickel toxicity—An overview. *J. Basic Clin. Physiol. Pharmacol.* **2018**, *30*, 141–152. [CrossRef] [PubMed]
41. Genchi, G.; Carocci, A.; Lauria, G.; Sinicropi, M.S.; Catalano, A. Nickel: Human Health and Environmental Toxicology. *Int. J. Environ. Res. Public Health* **2020**, *17*, 679. [CrossRef] [PubMed]

42. Tajchman, K.; Ukalska-Jaruga, A.; Ceacero, F.; Pecio, M.; Steiner-Bogdaszewska, Ż. Concentration of Macroelements and Trace Elements in Farmed Fallow Deer Antlers Depending on Age. *Animal* **2022**, *12*, 3409. [CrossRef]
43. Nowicka, W.; MacHoy, Z.; Gutowska, I.; Nocen, I.; Piotrowska, S.; Chlubek, D. Contents of Calcium, Magnesium, and Phosphorus in Antlers and Cranial Bones of the European Red Deer (*Cervus elaphus*) from Different Regions in Western Poland. *Pol. J. Environ Stud.* **2006**, *15*, 297–301.
44. Ofoe, R.; Thomas, R.H.; Asiedu, S.K.; Wang-Pruski, G.; Fofana, B.; Abbey, L. Aluminum in plant: Benefits, toxicity and tolerance mechanisms. *Front. Plant Sci.* **2023**, *13*, 1085998. [CrossRef]

animals

Article

Overview of Cyanide Poisoning in Cattle from *Sorghum halepense* and *S. bicolor* Cultivars in Northwest Italy

Stefano Giantin [1,†], Alberico Franzin [1,†], Fulvio Brusa [1], Vittoria Montemurro [1], Elena Bozzetta [1], Elisabetta Caprai [2], Giorgio Fedrizzi [2], Flavia Girolami [3] and Carlo Nebbia [3,*]

1 Istituto Zooprofilattico Sperimentale del Piemonte, Liguria e Valle d'Aosta (IZSPLV), Via Bologna 148, 10154 Turin, Italy; stefano.giantin@izsto.it (S.G.); alberico.franzin@izsto.it (A.F.); fulvio.brusa@izsto.it (F.B.); vittoria.montemurro@izsto.it (V.M.); elena.bozzetta@izsto.it (E.B.)

2 National Reference Laboratory for Natural Toxins in Food and Feed, Istituto Zooprofilattico Sperimentale della Lombardia e dell'Emilia Romagna (IZSLER), Via Pietro Fiorini 5, 40127 Bologna, Italy; elisabetta.caprai@izsler.it (E.C.); giorgio.fedrizzi@izsler.it (G.F.)

3 Department of Veterinary Sciences, University of Turin, Largo Paolo Braccini 2, 10095 Grugliasco, Italy; flavia.girolami@unito.it

* Correspondence: carlo.nebbia@unito.it

† These authors contributed equally to this work.

Simple Summary: Both wild (*Sorghum halepense*) and cultivated Sorghum (*Sorghum bicolor*) species are commonly used for animal feeding. However, sorghum plants naturally produce dhurrin, a cyanogenic glycoside releasing cyanide; this extremely toxic molecule potentially causes lethal toxicoses, particularly in ruminants. This was the case in a number of farms located in Northwest Italy during August 2022, under weather conditions (drought, tropical temperatures) known to increase plants' dhurrin content. Sixty-six bovines died after grazing Sorghum pastures (*Sorghum bicolor* or *Sorghum halepense*) or being fed with Sorghum-containing hay (*Sorghum halepense*). The reported clinico-pathological findings clearly indicated cyanide poisoning, and chemical analysis revealed high concentrations of dhurrin in the plant materials. The successful management of such toxicosis should rely on the prompt removal of the contaminated fodder and the administration of the antidote sodium thiosulphate. Dhurrin content should be carefully monitored, particularly in the hot season, and both farmers and veterinarians should be made aware of the risks associated with feeding cattle even with cultivated Sorghum, particularly if grown under adverse climatic conditions.

Abstract: *Sorghum* plants naturally produce dhurrin, a cyanogenic glycoside that may be hydrolysed to cyanide, resulting in often-lethal toxicoses. Ruminants are particularly sensitive to cyanogenic glycosides due to the active role of rumen microbiota in dhurrin hydrolysis. This work provides an overview of a poisoning outbreak that occurred in 5 farms in Northwest Italy in August 2022; a total of 66 cows died, and many others developed acute toxicosis after being fed on either cultivated (*Sorghum bicolor*) or wild Sorghum (*Sorghum halepense*). Clinical signs were recorded, and all cows received antidotal/supportive therapy. Dead animals were subjected to necropsy, and dhurrin content was determined in Sorghum specimens using an LC–MS/MS method. Rapid onset, severe respiratory distress, recumbency and convulsions were the main clinical features; bright red blood, a bitter almond smell and lung emphysema were consistently observed on necropsy. The combined i.v. and oral administration of sodium thiosulphate resulted in a rapid improvement of clinical signs. Dhurrin concentrations corresponding to cyanide levels higher than the tolerated threshold of 200 mg/kg were detected in sorghum specimens from 4 out of 5 involved farms; thereafter, such levels declined, reaching tolerable concentrations in September–October. Feeding cattle with wild or cultivated Sorghum as green fodder is a common practice in Northern Italy, especially in summer. However, care should be taken in case of adverse climatic conditions, such as severe drought and tropical temperatures (characterising summer 2022), which are reported to increase dhurrin synthesis and storage.

Keywords: *Sorghum bicolor; Sorghum halepense;* dhurrin; cyanide poisoning; cattle; drought

1. Introduction

Sorghum is a genus of plants belonging to the family of *Poaceae*, widely used as forage crop as well as human food and for biofuel production. Globally, the most cultivated species is *Sorghum bicolor* (L.) Moench, also known as broomcorn or great millet; this species is particularly widespread in the Americas and Africa, which, in 2017–2021, accounted together for more than 80% of total production. In 2021, Europe produced 1.9% of global Sorghum, with France being the highest-producing country at 386,040 t [1]. In Italy, the second highest-ranking European producer, the Sorghum yield reached 242,855 t in 2023, making it the fourth cereal after wheat, corn and rice [2]. Some wild species are also exploited for animal feeding, such as the widespread *Sorghum halepense* (L.) Pers., commonly referred to as Johnson grass, originating from the Mediterranean and Western Asia regions and now reported as an invasive weed across all continents [3].

When used as feed, Sorghum must be managed with particular caution because of a cyanogenic glycoside called dhurrin ((S)-4-Hydroxymandelonitrile β-D-glycoside) [4], which is synthesised as a secondary metabolite in its tissues. This molecule contains a cyanide group (CN^-) that can be released upon hydrolysis and is extremely toxic to all eukaryotic cells. CN^- inhibits cellular respiration by binding to the Fe^{+++} of cytochrome oxidase, rendering cells unable to utilise molecular oxygen and ultimately to synthetise ATP [5]. The rumen microbiota is able to rapidly hydrolyse dhurrin, further accelerating cyanide release and therefore making ruminants much more sensitive to CN^- than monogastric species [6,7]. Such rapid and massive CN^- release can cause severe, often lethal, poisonings in ruminants, particularly upon the ingestion of large amounts of fodder with high dhurrin content [6].

In Sorghum plants, dhurrin is produced especially during early growth phases [5,8]. Thanks to this glycoside, *Sorghum* species are quite resistant to herbivores, including insects [9]. Mature plants generally contain a lower amount of dhurrin and are therefore considered safe for animal feeding; however, dhurrin content is reported to increase under the following conditions [10]:

- Prolonged drought, frost, wilting, chewing and any other condition causing plant cell injury;
- Massive herbicide treatments;
- Extensive use of nitrogen-based fertilisers.

When used as feed, Sorghum can be directly grazed by animals or harvested for green forage, silage and hay production. Generally, the ensiling process leads to a dispersal of CN^- from plant tissues. Still, in some cases, high CN^- concentrations can remain in plants that have undergone rapid desiccation and subsequent conservation in large bales [9]. Because of the CN^- poisoning potential, Sorghum harvesting and use require cautious management in order to minimise poisoning risks, with special attention when used for feeding ruminants. Young leaves and new shoots, including the sprouts, are the most dangerous parts, as they can concentrate large amounts of dhurrin [6].

Although Sorghum toxicity has long been known, no poisoning cases in bovines have been reported in Europe in recent decades [11,12], with the exception of two cases in Spain quoted in a review on plant poisoning [13]. A search of the grey literature also revealed no results in Europe, but there were several cases in both the Americas and Australia [14]. Likewise, data on Sorghum poisonings were found—both in scientific databases and through online search engines—in extra-European countries, especially in semi-arid regions of South America [15–17] and India [18–20], where Sorghum cultivation for fodder purposes is common.

In August 2022, 66 bovines died in Piedmont (a region in Northwest Italy) after being exposed to the S. *bicolor* × *S. sudanense*—i.e., *S. bicolor* ssp. *sudanense* (P.) Stapf—cultivar called Suzy [21] or to forage containing *S. halepense*. The aim of this study is to provide a detailed overview of this outbreak, with special emphasis on the diagnosis and the therapeutic management of this toxicosis. Results of dhurrin concentration monitoring from August to November 2022 in both cultivated and wild *Sorghum* samples from the affected farms and elsewhere are also presented. A short preliminary report of the outbreak has been published in Italian in 2023 [22].

1.1. Poisoning Cases (August 2022)

Five outbreaks of Sorghum poisoning occurred in August 2022 in Piedmont. Figure 1 shows the epidemiological data concerning the poisoning cases.

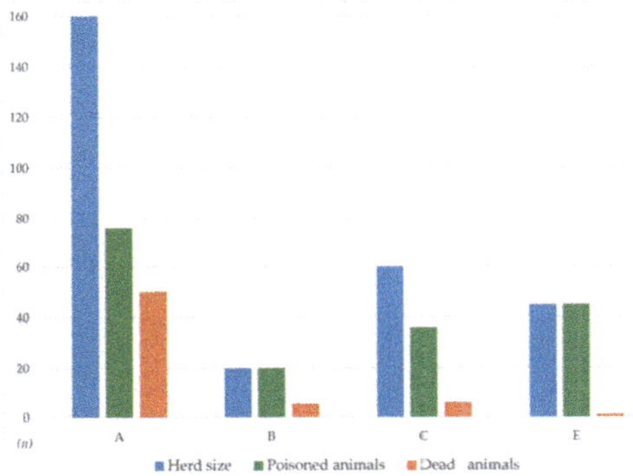

Figure 1. Epidemiological data (herd size, morbidity and mortality rate) of the five reported poisoning cases: A—Sommariva del Bosco; B—Moretta; C—Bra; E—Cossato. Case D (Asti) is not shown due to the lack of reliable information.

1.1.1. Case A—6th of August: Sommariva del Bosco (Cuneo)

A herd of 160 cows, mainly of the Piedmontese breed, was allowed free access to a field entirely cultivated with the *S. bicolor* × *S. sudanense* cultivar Suzy. As the animals were hungry due to overnight fasting, they rapidly ingested Sorghum plants, specifically sprouts with a height range of 30–45 cm. Around half of the animals were poisoned; forty-six of them rapidly died 20–30 min after the ingestion (Figure 2a), while in four further individuals, death ensued in the following hours. Most of the dead individuals were pregnant. The surviving ones were immediately moved away. Based on the clinical picture, the sudden deaths and the gross lesions (see below), cyanogenic glycoside poisoning was promptly suspected.

(a) (b)

Figure 2. Case A—Sommariva del Bosco (**a**) and B—Moretta (**b**): poisoned/dead animals in lateral recumbency, mostly on the right side.

1.1.2. Case B—11th of August: Moretta (Cuneo)

A group of 20 adult mixed-breed cows and bulls (mainly Friesian × Piedmontese or other meat breeds) housed in tie stalls were fed green chop (fresh forage) mainly composed of *S. halepense*. This is a common farming practice in Piedmont, especially during the warm season, when green and high-quality forage is scarce. All the affected animals were lactating cows aged more than 3 years; five of them suddenly died after being offered the contaminated feed (Figure 2b). The forage was promptly removed from the troughs after the onset of the clinical signs of poisoning.

1.1.3. Case C—11th of August: Bra (Cuneo)

Sixty adult cows of the Piedmontese breed housed in a free-stall barn were fed green chop, mainly consisting of Johnson grass. Poisoning signs were noticed during the subsequent night in thirty-six individuals; four of these suddenly died after grass ingestion, while in two further individuals, death ensued few days later. As in case B, the forage was removed after the first symptoms, and no further mortality was recorded.

1.1.4. Case D—12th of August: Asti

This case occurred in a cow–calf operation farm consisting of about 60 heads of the Piedmontese breed (cows and calves) housed in a free-stall barn. By day, the animals were allowed to graze on pastures in proximity to the farm for most of the year. Four cows died after ingestion of *S. halepense*, which was found to contaminate the pasture. This episode was tardily and poorly reported to the veterinarians, such that it was not possible to collect reliable epidemiologic information.

1.1.5. Case E—25th of August: Cossato (Biella)

The farm's characteristics were similar to those from case D, i.e., a cow–calf operation farm with about 45 head, mainly of the Piedmontese breed (but also meat crossbreds). For most of the year, animals were free to graze on pastures surrounding the farm. All cows showed the typical signs of cyanide poisoning, mainly respiratory distress and a tendency toward recumbency; overall, symptoms were less severe than in cases A, B and C, resulting in the loss of only one cow. Also in this case, the cause of poisoning was pasture contamination with *S. halepense*.

2. Materials and Methods

2.1. Necropsies and Histological Analysis

Due to unfavourable conditions (high external temperatures and the limited availability of veterinarians), necropsies were performed on only a few animals ($n = 6$ in total) directly at the farms. Heart, lung, brain, liver, kidney, spleen, reticulum, rumen, omasum, abomasum and intestine samples were collected, fixed in 10% buffered formalin (4% formaldehyde), dehydrated and embedded in paraffin wax blocks. Each sample was then sectioned at 4–5 μm-thickness, mounted on glass slides and stained with haematoxylin and eosin to reveal histopathological alterations. Slides were examined by two independent veterinary pathologists.

2.2. Sorghum Sample Collection

To confirm the suspicion of cyanogenic glycoside poisoning, samples of Sorghum to which the cattle were exposed were collected at each farm involved in the outbreak (Figure 3) and submitted for dhurrin determination (see below). It was also decided to collect and analyse additional specimens of both wild and cultivated Sorghum in order to measure dhurrin content in plants from different areas of the Piedmont region. In particular, the selection process was based on three main factors:

- Areas where strong drought conditions occurred [23];
- Requests for dhurrin analysis from a number of worried farmers;
- Financial resources.

Figure 3. Map of the Piedmont region showing the locations of the poisoning cases (blue triangles: A—Sommariva del Bosco, B—Moretta, C—Bra, D—Asti, E—Cossato) and the other farms selected for sampling of either cultivated Sorghum (red circle: M—Montechiaro d'Asti) or wild Sorghum (red circles: 1—Cuneo, 2—Faule, 3—Montiglio Monferrato, 4—Sampeyre, 5—Sanfrè, 6—Verrua Savoia). All the samples (n = 57) were collected from August to November 2022. Orange lines indicate province borders.

In addition, in certain instances, samples were collected from different plant portions and at diverse growth stages.

As regards cultivated Sorghum, the *S. bicolor* × *S. sudanense* cultivar Suzy was involved in case A. Two cultivars, *S. bicolor* ssp. *drummondii* Piper and *S. bicolor* × *S. sudanense* Sudal [24], were then sampled from a farm in Montechiaro d'Asti (Asti), which was experiencing similar drought conditions to the farm in case A; Sorghum had not yet been harvested due to the severe outbreak that had occurred in Sommariva del Bosco.

Common Johnson grass, which is frequently used as fodder by Piedmontese farmers, was the cause of poisoning cases B, C, D and E. Further sites for *S. halepense* sampling were Verrua Savoia (Torino province), Montiglio Monferrato (Asti province), Cuneo, Faule, Sampeyre and Sanfrè (Cuneo province).

One pooled sample composed of a minimum of 500 g of fresh plant materials was collected randomly from different areas inside pasture fields or directly taken from the green forage offered to animals in the stalls. Additionally, in one case (A), the rumen content was collected from a dead cow. All sampling activities were completed from August to November 2022.

2.3. Dhurrin Determination

Samples were analysed using an in-house liquid chromatography–tandem mass spectrometry (LC–MS/MS) method at the National Reference Laboratory for Plant Toxins, Food Chemical Department of Istituto Zooprofilattico Sperimentale della Lombardia e dell'Emilia Romagna (IZSLER), located in Bologna. Samples were ground into flour, and 1 ± 0.1 g of each one was extracted with 6 mL of aqueous methanol (80%). The sample was shaken vigorously for 30 s and placed in an ultrasonic water bath for 15 min. The mixture was centrifugated for 5 min at 4000× *g*, and the supernatant was collected in another tube. This extraction was repeated twice, and the supernatant was combined and made up to a volume of 20 mL with water. Thereafter, 1 mL of solution was evaporated to dryness under a stream of nitrogen at 40 °C. The residue was dissolved in 0.5 mL of 10% methanol in aqueous solution, diluted and analysed by using LC–MS/MS.

The LC–MS/MS analysis was performed on a XEVO Tq-XS Acquity ultra-performance liquid chromatograph (UPLC) I Class Plus from Waters (Milford, MA, USA). Chromatographic separation was achieved on an Acquity UPLC C8 BEH column measuring 100 mm × 2.1 mm, 1.7 µm (Water Corporation, Milford, MA, USA). Data acquisition and processing were carried out using MassLynx software v. 4.2. SCN1012. Mobile phase A consisted of 0.1% formic acid in water/acetonitrile (95:5, *v*/*v*), and mobile phase B consisted of 0.1% formic acid in acetonitrile. The following gradient was used: 0–0.5 min, isocratic 2% B; 0.5–4 min, linear gradient 2–50% B; return to initial conditions in 0.5 min and hold for 1 min. The total run time was 6 min. The flow rate was 0.4 mL/min. The injection volume was set at 5 µL. The ESI source operated in positive ionisation mode with the following instrumental parameters: capillary voltage of 0.5 kV, cone voltage of 40 V, source temperature of 120 °C and desolvation temperature of 600 °C. The conditions of ionisation and fragmentation were identified by continuous infusion of tuning solutions and gradual adjustment of the parameters. According to SANTE/12089/2016 [25], dhurrin was identified by the retention time, ion fragments and ion ratio. LC–MS/MS parameters for dhurrin determination (retention time, precursor ions, daughter ions and fragmentation conditions) are shown in Table S1. The retention time was within ±0.2 min of the reference peaks. The peaks showed similar shapes and overlapped with each other. The ion ratio was within ±30% of the average of the calibration standards from the same sequence. The peaks were within the linear range of the detector with an $S/N \geq 3$ [26]. The LC–MS/MS method's selectivity was evaluated by acquiring the data in MRM mode and monitoring one precursor ion and two daughter ions for each molecule [25].

A multi-level calibration curve with concentration levels from lowest to highest (0.2–0.5–1–2.5–5–10–15 µg/mL) was prepared in 10% methanol in aqueous solution. A correlation coefficient (R^2) ≥ 0.99 and a normal distribution of residuals lower than 20% were achieved in every analytical batch. The calibration curve, a representative chromatogram of dhurrin reference material (2.5 µg/mL) and a chromatogram of a blank and a contaminated Sorghum sample are shown in Figures S1–S4.

The limit of quantification (LOQ) of dhurrin in feed was 50 mg/kg, corresponding to 4.3 mg/kg hydrogen cyanide (HCN), i.e., cyanide. It has been evaluated under conditions of accuracy and precision, verifying the signal-to-noise ratio to be at least equal to 10. The recovery % (70–120) of the quality control spiked at LOQ was in line with the guidance document on performance criteria of the European Union Reference Laboratory for Mycotoxin and Plant Toxins [26]. According to EFSA [27], 1 g of dhurrin has an HCN potential of 86.7 mg, representing the total amount of HCN released under conditions of complete hydrolysis of the present dhurrin. For the sake of simplicity, in this paper, the HCN potential is referred to as HCN/cyanide concentration.

2.4. Clinical Picture

Poisoned bovines showed multiple symptoms, with variable distribution among individuals. Many cows were found in sternal or lateral recumbency, mainly on the right side. Respiratory distress was observed in most of the poisoned animals, consisting of tachypnoea, dyspnoea, panting and gasping. Several cows also displayed stupor, convulsions and muscle twitching with vocalisations (mooing). Sialorrhoea was an additional common symptom among poisoned bovines. Moreover, light to moderate tympanism was detected in a few individuals. Hyperthermia, nystagmus, mydriasis and wheezes were occasionally observed.

2.5. Therapeutical Protocols

Table 1 depicts the treatment performed in each case and the relative success rate.

Table 1. Treatments given, n. of surviving animals and therapeutical success rate in the described outbreak of Sorghum toxicosis. Case D is not included because cows were not subjected to any treatment.

Case	Type of Treatment	Treated Animals	Surviving Animals	Success Rate
A	Rehydrating solutions + multivitamin complex	30	26	87%
B	Sodium thiosulphate	5	5	100%
C	Methylene blue and (later) sodium thiosulphate	30	28	93%
E	Sodium thiosulphate	40	40	100%

2.5.1. Case A

Although, as mentioned above, a cyanogenic glycoside poisoning was suspected, it was difficult to find the proper remedies also because this outbreak happened during the weekend. Thirty animals were treated intravenously with a mix of rehydrating solutions (Ringer's lactate, physiological and glucose solutions), coupled with 60 mL of the multivitamin Dobetin B1® (cyanocobalamin 1 mg/mL, thiamine hydrochloride 100 mg/mL). Considered the hot external temperature (over 38 °C), the cows were also cooled down by spraying with water taken from the mobile drinking troughs. Twenty-six of the treated animals survived.

2.5.2. Case B

Owing to the similarity to the clinical picture described for the Sommariva del Bosco poisoning (case A) and based on the first analytical results revealing the mass presence of dhurrin in sorghum samples from that case, antidotal therapy was immediately started. However, due to the limited availability of sodium thiosulphate ($Na_2S_2O_3$), it was decided to treat only the most severely affected individuals (n = 5), lying in sternal/lateral recumbency with panting and vocalisations. Antidote solution was prepared by dissolving 5 g $Na_2S_2O_3$ in 4 L of Ringer's lactate, which was slowly administered i.v. (Figure 4).

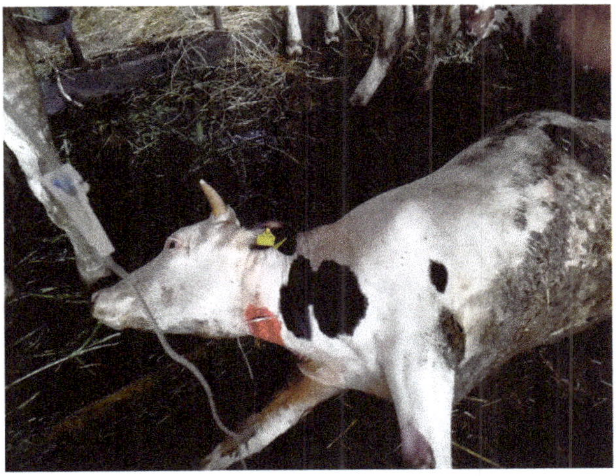

Figure 4. In Moretta (case B), a poisoned cow receiving the antidote (sodium thiosulphate) i.v. Note the cherry-red blood on the neck of the animal.

Furthermore, 15 g of $Na_2S_2O_3$ was dissolved in 10 L of cold water and then given orally through drench guns. At 10–15 min after antidote administration, breathing started to improve, and vocalisations almost ceased; the cows were again able to stand in about one hour.

2.5.3. Case C

As mentioned above, poisoning symptoms were noticed during the night, and this resulted in difficulties in obtaining $Na_2S_2O_3$ in sufficient amounts to treat all the affected animals (n = 30). It was therefore decided to administer methylene blue i.v. (10 g dissolved in 4 L of Ringer's lactate) first; however, this treatment was only partially effective in reducing the severity of the clinical signs. As soon as $Na_2S_2O_3$ was fully available (late in the morning), it was promptly administered i.v. (5 g dissolved in 4 L rehydrating solution) to all previously treated cows. This led to a rapid improvement of the clinical picture as described for case B. Twenty-eight cows survived, while two died few days later.

2.5.4. Case D

No treatment was performed.

2.5.5. Case E

Due to the alert system set up for the purpose of tackling the cyanogenic glycoside outbreaks, the antidote $Na_2S_2O_3$ was made readily available to veterinarians. Accordingly, all poisoned animals were treated with the antidote as soon as 1 h after the onset of clinical signs, and a rapid recovery ensued within 2 h from therapeutic intervention. The treatment schedule was the one detailed for case B.

3. Results

3.1. *Gross Lesions*

Necropsy was performed on six carcasses: three from case A, one from case B and two from case C.

On post mortem examination, an intense sweet smell of bitter almonds was reported. The blood was bright red and clotted poorly. The tracheas were congested, with oedema, petechiae and a variable amount of froth; severe pulmonary emphysema and oedema were also noticed (Figure 5a). Hydropericardium, focal haemorrhages and necrosis of the myocardium were observed (Figure 5b). The rumens were filled with fresh green material and bloated; suffusion and petechiae were present on the rumens, reticula and omasa as well. Congestion and petechial haemorrhages were observed in the gastrointestinal tracts. Finally, abomasitis (Figure 5c), severe splenomegaly, and enlarged and congested livers were observed in most of the animals. In a carcass from case C belonging to a cow that underwent the antidotal therapy but died two days after treatment, subcutaneous gelatinous necrosis was additionally detected.

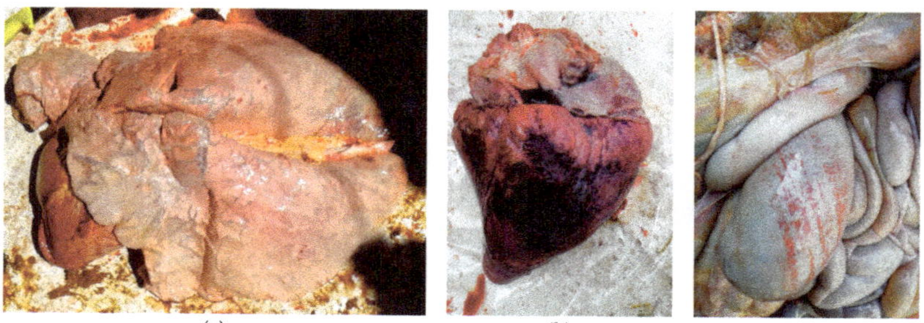

(a) (b) (c)

Figure 5. Necropsy findings in Moretta case (B), revealing lung emphysema (**a**), myocardial haemorrhages (**b**) and abomasitis (**c**).

3.2. Histopathological Lesions

In most of the animals, the pulmonary parenchyma revealed foci of alveolar oedema, emphysema and congestion of capillaries. Histologically, type I pneumocyte necrosis and hyperplasia of type II pneumocytes with hyaline membranes were observed, along with thickening of septa due to mononuclear cell infiltration. Three individuals presented hearts with large haemorrhagic areas in myocardial and pericardial adipose tissue. Focal fibrosis and moderate multifocal non-purulent myocarditis were also observed.

In one cow, which died two days after treatment (case C), the abomasum revealed acute abomasitis characterised by severe hyperaemia, red–brown haemorrhagic exudate adherent to the mucosa and neutrophilic infiltration with focal oedematous–haemorrhagic fluid in the submucosa. Thrombosis and regressive epithelial alterations were also observed in the mucosa.

3.3. Dhurrin Determination

Sorghum samples were taken from the farms involved in poisoning outbreaks and from other selected farms and fields, as detailed in Materials and Methods.

Table 2 reports the results related to the five intoxication cases. In case D, two samples from different areas of the pasture were collected. In case E, three samples from distinct parts of the plants and stages of maturity were tested. HCN concentrations are expressed in mg/kg; values over 200 mg/kg are generally considered dangerous [6,28]. In all but one case, Sorghum HCN concentrations were higher than 200 mg/kg; only in case E were there no samples that reached 200 mg/kg. Additionally, in one sample picked on the border of the pasture from case D, the HCN concentration measured 9 mg/kg.

Table 2. Dhurrin and hydrogen cyanide (HCN) concentrations in *Sorghum* samples implied in the five outbreaks of cyanogenic glycoside poisoning occurring in Piedmont in August 2022. When plant part is not specified, analysis was performed on the whole plant.

Case	Date	Sorghum Species	Site of Collection	Dhurrin (mg/kg)	HCN (mg/kg)
A	6-Aug	Suzy [1] (sprout, height 30–45 cm)	Pasture	10.717	929
B	16-Aug	*S. halepense*	Trough	5.627	487
C	16-Aug	*S. halepense*	Trough	7.961	690
D	16-Aug	*S. halepense*	Pasture	4.834	419
		S. halepense	Pasture border	104	9
E	25-Aug	*S. halepense* (young plants)	Pasture	335	29
		S. halepense (leaf mix)	Pasture	488	42
		S. halepense (inflorescence)	Pasture	105	9

[1] *S. bicolor* × *S. sudanense* cultivar called Suzy.

One sample of ruminal content from a dead cow (case A) was also tested for dhurrin content with a negative result.

Dhurrin and HCN concentrations from the two farms selected for cultivated forage Sorghum are reported in Table 3 (Case A—Sommariva del Bosco) and Table 4 (Montechiaro d'Asti). For both farms, HCN content remained high for the whole period of August, tended to decline toward tolerable levels in September and reached negligible levels only in fall.

Taking the results together, no clear differences in HCN concentrations between different portions of the plants were noticed, with the exception of the culm. Indeed, in Montechiaro d'Asti, two culm samples collected on the 30th of August revealed lower HCN concentrations (0 and 147 mg/kg, respectively) than leaves and inflorescences (n = 7), which displayed HCN values in the range of 155–868 mg/kg. additionally, no strong evidence of a higher HCN content was found in younger/shorter individuals with respect to older/taller ones, even though two samples picked from ensiled bales on the 17th of

November revealed a low HCN level (19 mg/kg) in a bale made of short immature plants (height < 50 cm) and no HCN at all in a bale made of mature individuals (height > 150 cm).

Table 3. Time course of dhurrin and hydrogen cyanide (HCN) concentrations in the *Sorghum bicolor* × *Sorghum sudanense* variety called Suzy from Sommariva del Bosco (case A). When plant part is not specified, analysis was performed on the whole plant.

Date	Dhurrin (mg/kg)	HCN (mg/kg)	Notes
6-Aug	10,717	929	Sample related to the outbreak (case A)
14-Aug	6869	596	Average plant height 50 cm
16-Aug	14,246	1235	Open-air dried
16-Aug	5590	485	Fresh, leaves > 1 m
16-Aug	8300	720	Fresh, leaves < 50 cm
17-Aug	<LOQ	0	Bundled; cut on the 14th of July
21-Aug	6550	568	Average plant height 60 cm
27-Aug	7661	664	Average plant height 68 cm
5-Sept	1420	123	
12-Sept	1798	155	
23-Sept	958	83	
27-Sept	974	84	Field "Paolorio"
27-Sept	1354	117	Field "Luppiano"
27-Sept	2707	235	Field "Valè"
6-Oct [1]	<LOQ	0	Fresh, chopped
23-Nov	<LOQ	0	Mature silage (45 days), from mixed fields

[1] The same result (0 mg/kg HCN) was measured in six samples from 5 different fields.

Table 4. Time course of dhurrin and hydrogen cyanide (HCN) concentrations in *Sorghum* samples from a farm in the Asti province (Montechiaro d'Asti), located near case D. All the samples belong to mixed individuals grown from a seed mixture of two varieties: *S. bicolor* ssp. *drummondii* Piper and *S. bicolor* × *S. sudanense* Sudal. When plant part is not specified, analysis was performed on the whole plant.

Date	Dhurrin (mg/kg)	HCN (mg/kg)	Notes
16-Aug	9770	847	Sowed at the beginning of June, never cut
16-Aug	2840	246	Grown-back plants
16-Aug	2065	179	Sowed at the beginning of June, grazed in July
30-Aug	1792	155	Field 1; leaves > 150 cm
30-Aug	1919	166	Field 1; leaves ~ 50 cm
30-Aug	<LOQ	0	Field 1; culm
30-Aug	3251	282	Field 1; inflorescence
30-Aug	1865	162	Field 1; grown-back plants, without roots
30-Aug	10,010	868	Field 2; leaves > 150 cm
30-Aug	6701	581	Field 2; leaves ~ 50 cm
30-Aug	1697	147	Field 2; culm
30-Aug	4553	395	Field 2; inflorescence
26-Sept	3967	344	Culm and leaves
26-Sept	205	18	Inflorescence
17-Nov	229	19	Immature plants (without grains) ~ 50 cm; from ensiled bale
17-Nov	<LOQ	0	Mature plants (with grains) > 150 cm; from ensiled bale

Concurrently, a set of samples of *S. halepense* were gathered from several farms and fields scattered across Cuneo, Asti and Torino provinces. Their dhurrin and HCN concentrations are listed in Table 5. Despite the close proximity of some of the sampling sites to farms experiencing poisoning cases from cyanogenic glycosides, only in one case was the threshold of concern of 200 mg/kg HCN reached, even in specimens collected in August.

The frost-covered sample collected on the 23rd of November in Sampeyre, in a mountain area, revealed dhurrin concentrations <LOQ. In addition, a negative result was found in a sample from case D made of mixed grasses.

Table 5. Dhurrin and hydrogen cyanide (HCN) concentrations in *Sorghum halepense* collected from different farms and fields in Piedmont during 2022. The analysis was performed on the whole plants.

Date	Location	Province	Dhurrin (mg/kg)	HCN (mg/kg)	Notes
16-Aug	Asti	Asti	<LOQ	0	Mixed grasses
17-Aug	Faule	Cuneo	85	7	Cut for haymaking
17-Aug	Verrua Savoia	Torino	1558	135	Field used for haymaking
17-Aug	Montiglio M.to	Asti	2036	176	Field "Sant'Anna"
17-Aug	Montiglio M.to	Asti	2693	233	Field "Acquedotto"
17-Aug	Montiglio M.to	Asti	1917	166	Field "Vallone"
5-Sept	Bra	Cuneo	289	25	Field of case C; forage for silo
15-Sept	Cuneo	Cuneo	57	5	Plants > 50 cm
15-Sept	Cuneo	Cuneo	401	35	Plants < 50 cm
16-Sept	Sanfrè	Cuneo	<LOQ	0	Mature plants (with inflorescence)
23-Nov	Sampeyre	Cuneo	<LOQ	0	Frost-covered plants

The seasonal trend of cyanide concentrations in all collected samples (n = 57) of either cultivated or wild Sorghum is depicted in Figure 6. Overall, a clear decreasing trend was noted: in August 2022, 58% of samples were found to contain levels > 200 mg/kg, whereas, from September to November, such amounts were detected in only 8% of the specimens.

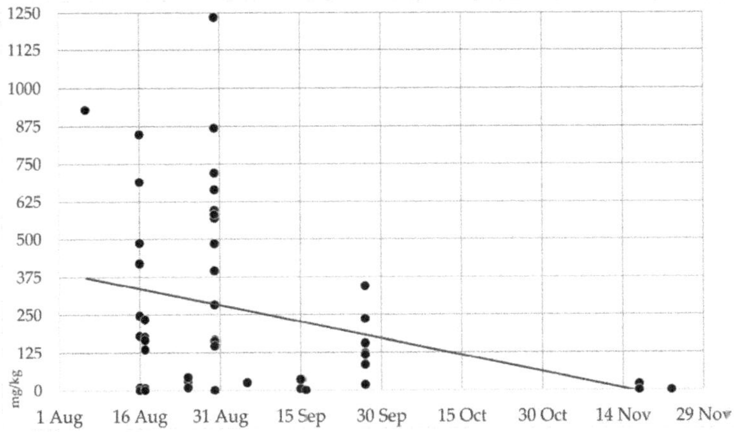

Figure 6. Seasonal trend of hydrogen cyanide (HCN) concentrations in all cultivated and wild collected Sorghum samples. Figure shows aggregate data (n = 57) from all samples collected in Piedmont, including all areas, from August to November 2022.

4. Discussion

The rapid onset of clinical signs in cows shortly after the ingestion of Sorghum, followed sometimes by sudden death, had immediately suggested cyanide poisoning. Respiratory distress, stupor, sternal or lateral recumbency, convulsions, muscle tremors and sialorrhoea are typically reported in cyanide poisoning in cattle [6,29]. In addition, the recorded intense sweet odour of "bitter almonds", the bright cherry-red colour of venous blood and the presence of lung congestion and emphysema as well as the presence of froth in the trachea are consistently recorded in bovines with cyanogenic glycoside poisoning [30]. The detection of abomasitis that features oedematous–haemorrhagic and neutrophilic granulocyte infiltrations has been also associated with cyanide poisoning [6]. Finally, myocardial haemorrhages further point to cyanide poisoning [31].

The gold-standard therapy for cyanide toxicosis [6,32] consists of supplying a chemical agent able to induce the formation of methaemoglobin (MetHb), i.e., oxidised (Fe^{+++}) haemoglobin, which is unable to bind O_2, and making it available to tissues. However, cyanide shows a higher affinity toward the Fe^{+++} central haem iron of MetHb than the Fe^{+++} of cytochrome oxidase. This causes the release of cyanide from the enzyme, the formation of cyanoMetHb and the reactivation of cell respiration. MetHb formation in large animals may be primarily accomplished by administering sodium nitrite i.v. (10 to 20 mg/kg bw); this treatment should be repeated with great care because of the danger of producing nitrite toxicosis, with further impairment of cellular respiration and severe hypotension [30]. Methylene blue at high dosages (1 to 3 g/~250 kg bw) has been recommended as an alternative to nitrites [31]. This treatment must be coupled with the sulphur donor $Na_2S_2O_3$, which, in the presence of rhodanese, reacts with HCN, yielding thiocyanate (SCN^-); this metabolite lacks any detrimental effects on cellular respiration and is rapidly excreted via the kidneys. In the reported cases herein, coupling methylene blue and $Na_2S_2O_3$ administration did not seem to result in a visible improvement in therapeutic efficacy; a significant and rapid relief of the clinical signs was indeed obtained only after $Na_2S_2O_3$ treatment, which was successfully used alone in cases B and E with 100% efficacy. It has actually been reported that, in cattle, there is no benefit of administering i.v. a MetHb-inducing agent over $Na_2S_2O_3$ alone [32]. In addition, prompt oral dosing with $Na_2S_2O_3$ may help in detoxifying HCN released in the rumen even before the onset of clinical signs [33]. The overall good success of the antidotal treatment further confirmed the diagnosis of cyanide poisoning. It should be noted that treated cows from case A had a relatively high survival rate (87%) even though they did not receive specific antidotes, but only a palliative fluid therapy with a multivitamin complex. The prompt removal of the animals from the contaminated pasture, i.e., after the first sudden deaths, was likely the cause of the high recovery rate.

According to the European Directive 2002/32/EC [34], a threshold of 50 mg/kg cyanide has been established for animal feed and raw materials. Under field conditions, concentrations over 200 mg/kg are considered sufficient to induce overt toxicosis [6,28,31].

It is generally assumed that crop plants are less resistant to parasites and herbivores than their wild counterparts due to artificial genetic selection aiming at reducing the content of specific defence compounds (e.g., cyanogenic glycosides) that may prove harmful for humans and livestock [35]. However, this assumption cannot be generalised to Sorghum. Unexpectedly, broomcorn cultivars such as Suzy (a *S. bicolor* × *S. sudanense* variety, Sommariva del Bosco, case A) and the mixture of Piper and Sudal (Montechiaro d'Asti), revealed very high HCN concentrations in August 2022. Both cultivars are specifically marketed for animal feeding purposes; however, guidelines for use reported on seeds' envelopes do recommend not to feed the crop to animals when plants are below 70/80 cm (70 cm for the mixture of Piper and Sudal, and 80 cm for Suzy), but they lack any information on the potential related danger [36]. In case A, the farmer decided to allow his herd to graze on the field although the sorghum plants were below the recommended height. As was true for many other farmers during that summer, his farm was experiencing a shortage of forage due to its high cost and the scarce availability of green pastures. The increase in forage prices was a direct consequence of a lower supply on the market that, in turn, was caused by a diffuse drought particularly affecting Northwest Italy. A parallel survey was conducted on cultivated hybrids (*S. bicolor* ssp. *Drummondii* Piper and *S. bicolor* × *S. sudanense* Sudal) from different fields surrounding a farm in the Asti province (Montechiaro d'Asti) near poisoning case D; HCN concentrations >200 mg/kg were detected in 50% of specimens collected in August 2022, with peaks of 847–868 mg/kg. Overall, our findings confirm that bovines should not be fed on young plants even of cultivated hybrids, including regrowth after cutting, because of the high risk of cyanide poisoning.

In the outbreak of cyanogenic glycoside poisoning in cows described herein, *S. halepense* was implicated in 4 out 5 cases. Johnson grass is considered among the most invasive and dangerous weeds in Europe and extra-European countries; aside from the potential accumulation of toxic amounts of cyanogenic glycosides, several potentially adverse effects have been reported, including displacement of natural flora; competition with other crops; synthesis of allelochemicals interfering with crop growth; and hosting of plant pathogens (for a review, see Peerzada et al., 2017 [37], and the numerous literature references therein). Despite that, the free growth of Johnson grass is rarely counteracted; in fact, as reported in four cases (B to E), farmers traditionally employ Johnson grass as a fodder plant (hay or pasture) during periods of droughts. As with other Sorghum species, several factors, including soil chemical composition, plant age, use of nitrogen fertilisers, weather conditions and damage to plant tissues, are reported to affect the dhurrin content and hence the potential HCN release of Johnson grass [34]. There is scant information on the dhurrin and HCN content of *S. halepense*, particularly from European countries. In a study performed in India, calculated HCN concentrations (based on the colorimetric method) of uncultivated Johnson grass from farm bunds averaged around 900 mg/kg at 30 days after weeding but fell to 120 mg/kg at the 25% flowering stage [38]. Therefore, as with cultivated Sorghum species, cattle should not be fed with Johnson grass at the early stage of the crop. In the outbreaks reported here, three poisoning cases concerning *S. halepense* revealed HCN concentrations in the range 419–690 mg/kg (cases B, C, D). The relatively low amount of HCN (below 50 mg/kg) detected in plant specimens from case E is probably attributable to uncorrected sampling procedures. For comparison, samples of *S. halepense* were collected in a more scattered way during August and September 2022 in fields from farms located in different areas of Piedmont, even near poisoning cases (Sanfrè, Faule, Montiglio Monferrato); of note, only in one case were HCN amounts >200 mg/kg detected in plant specimens, likely pointing to the occurrence of different pedo-climatic conditions not resulting in remarkable accumulation of dhurrin as was reported for the areas of the outbreak.

As a matter of fact, in summer 2022, unfavourable weather conditions were registered all across Europe, and Northern Italy, particularly certain areas of Piedmont, resulted one of the driest regions [39]. According to the Piedmont Regional Agency for Environmental Protection (ARPA), summer 2022 was one of the hottest and driest of the last 30 years in Piedmont [23]. Indeed, during that summer, unprecedented temperatures were registered, occasionally reaching all-time highs (Figure S5). Additionally, the numbers of tropical nights (T > 20 °C) and days (T > 30 °C) were higher than in previous years (Table S2). Moreover, rainfalls were irregular in terms of both quantity and regional distribution, with a decrease of 50–60% with respect to previous years, especially in areas where cyanide poisoning outbreaks occurred (Figures S6 and S7). Finally, the hydric balance had been in deficit since the previous winter (Figure S8), also due to limited snow reserves. These conditions are reasonably believed to be responsible for the excessive accumulation of dhurrin observed in most *Sorghum* specimens collected in the outbreak area and surrounding areas.

5. Conclusions

The use of *Sorghum* plants for cattle feeding is a common practice in Northern Italy, where the most abundant species are the wild weed *S. halepense* and several cultivars of *S. bicolor*. Data from the described outbreaks in Piedmont indicate that not only Johnson grass but also different Sorghum cultivars may accumulate dhurrin concentrations that can trigger lethal poisoning of cows, particularly if animals are fed with young plants or sprouts and under adverse climatic conditions (drought, tropical temperatures). As expected, the decrease in diurnal and nocturnal temperatures together with the increase in the amount of precipitation occurring in October and November caused dhurrin levels to decline and, thus, HCN concentrations to fall well below the toxic threshold of 200 mg/kg.

Accurate management is needed when using Sorghum forages, and specific instructions should be reported on every commercial *Sorghum* seed envelope. Moreover, both farmer unions and local authorities should disseminate technical information on how to avoid toxicoses when feeding animals with cultivated Sorghum, including instructions/good practices for the safe use of *S. halepense* as a fodder. Finally, our data further support the use of sodium thiosulphate alone in the treatment of cyanogenic glycoside poisonings in cattle, suggesting that this antidote should be made readily available to veterinary practitioners in order to ensure a rapid and efficacious intervention.

Supplementary Materials: The following supporting information can be downloaded at: https://www.mdpi.com/article/10.3390/ani14050743/s1, Table S1: LC–MS/MS parameters for dhurrin determination; Table S2: Numbers of tropical days (T > 30 °C) and nights (T > 20 °C) measured in every province of Piedmont in 2022; Figure S1: Calibration curve of dhurrin; Figure S2: Representative chromatograms of dhurrin reference material at 2.5 µg/mL; Figure S3: Representative chromatograms of a Sorghum sample; Figure S4: Representative chromatograms of a blank Sorghum sample; Figure S5: Daily temperature anomaly in Piedmont during the summer 2022 compared to the period 1991–2020; Figure S6: Rainfall anomaly (%) in Piedmont basins during summer 2022 compared to the period 1991–2020; Figure S7: Rainfall anomaly (%) in Piedmont basins during September 2022 compared to the period 1991–2020; Figure S8: Daily hydro-climatic balance (i.e., the difference between rainfall and evapotranspiration expressed in mm) in Piedmont in 2022 compared to the period 1959–2022. Data in Table S2 and Figures S5–S8 were extracted from the 2022 drought report, published by the Piedmont Regional Agency for Environmental Protection (ARPA) [23], openly available at https://www.arpa.piemonte.it/news/pubblicato-il-rappporto-sulla-siccita2019-in-piemonte-nel-2022 (accessed on 22 November 2023).

Author Contributions: Formal analysis: A.F.; funding acquisition: F.B.; investigation: S.G., A.F., V.M., E.B., E.C. and G.F.; resources: S.G., A.F. and F.B.; supervision: C.N.; visualisation: A.F. and F.G.; writing—original draft preparation: A.F., V.M., E.C., E.B., G.F. and S.G.; writing—review and editing: A.F., F.G. and C.N. All authors have read and agreed to the published version of the manuscript.

Funding: This study was financially supported from *Fondazione CRT*, a private non-profit organisation (Turin, Italy; grant number 2022.1784 "Il pascolo e la tossicità delle piante erbacee: prevenzione e trattamento", 7/12/2022).

Institutional Review Board Statement: Not applicable.

Data Availability Statement: Data are contained within the article.

Acknowledgments: The authors would like to thank, for their availability during sampling and data collection, the farmers G. Olivero (Sommariva del Bosco, case A), E. Bollati (Moretta, case B), M. Cravero (Bra, case C), D. Viarengo (Asti, case D) and M. Bandiera (Montechiaro d'Asti); the authors also thank the veterinarians who treated the cows (G. Topi–solo practitioner F. Barberis and F. Acciardi of ASL CN 2–Veterinary Services Cuneo province).

Conflicts of Interest: The authors declare no conflict of interest. The funders had no role in the design of the study; in the collection, analyses or interpretation of data; in the writing of the manuscript; or in the decision to publish the results.

References

1. FAOSTAT. Available online: www.fao.org/faostat/en/#data/QCL (accessed on 22 November 2023).
2. ISTAT. Available online: http://dati.istat.it/Index.aspx?DataSetCode=DCSP_COLTIVAZIONI (accessed on 21 December 2023).
3. Gunes, E.; Uludag, A.; Uremis, I. Econ Supplementary Material citationomic impact of johnsongrass (*Sorghum halepense* [L.] Pers.) in cotton production in Turkey. *J. Plant Dis. Prot.* **2008**, *21*, 515–520.
4. Yadav, M.; Singh, I.K.; Singh, A. Dhurrin: A naturally occurring phytochemical as a weapon against insect herbivores. *Phytochemistry* **2023**, *205*, 113483. [CrossRef] [PubMed]
5. Gleadow, R.M.; Møller, B.L. Cyanogenic glycosides: Synthesis, physiology, and phenotypic plasticity. *Annu. Rev. Plant Biol.* **2014**, *65*, 155–185. [CrossRef] [PubMed]
6. Panter, K.E. Cyanogenic Glycoside–Containing Plants. In *Veterinary Toxicology: Basic and Clinical Principles*, 3rd ed.; Gupta, R.C., Ed.; Elsevier: Amsterdam, The Netherlands; Academic Press: Cambridge, MA, USA, 2018; pp. 935–940.
7. Bhat, B.V. Breeding Forage Sorghum. In *Breeding Sorghum for Diverse End Uses*, 1st ed.; Aruna, C., Visarada, K.B.R.S., Venkatesh Bhat, B., Vilas Tonapi, A., Eds.; Woodhead Publishing: Cambridge, UK, 2018; pp. 175–191.

8. Mandrone, M.; Chiocchio, I.; Barbanti, L.; Tomasi, P.; Tacchini, M.; Poli, F. Metabolomic Study of Sorghum (*Sorghum bicolor*) to Interpret Plant Behavior under Variable Field Conditions in View of Smart Agriculture Applications *J. Agric. Food Chem.* **2021**, *69*, 1132–1145. [CrossRef] [PubMed]
9. Nahrstedt, A. Cyanogenic compounds as protecting agent for organisms. *Plant Syst. Evol.* **1985**, *150*, 35–47. [CrossRef]
10. Ouma, L.A.; Cheruiyot, E.K.; Ogendo, J.O. Biosynthesis and Role of Dhurrin in Forage Sorghum. *Rev. Agric. Sci.* **2023**, *11*, 259–270. [CrossRef] [PubMed]
11. Bertero, A.; Davanzo, F.; Rivolta, M.; Cortinovis, C.; Vasquez, A.; Le Mura, A.; Masuelli, A.; Caloni, F. Plants and zootoxins: Toxico-epidemiological investigation in domestic animals. *Toxicon* **2021**, *196*, 25–31. [CrossRef] [PubMed]
12. Nagy, A.-L.; Ardelean, S.; Chapuis, R.J.J.; Bouillon, J.; Pivariu, D.; Dreanca, A.I.; Caloni, F. Emerging Plant Intoxications in Domestic Animals: A European Perspective. *Toxins* **2023**, *15*, 442. [CrossRef]
13. García-Arroyo, R.; Quiles Sotillo, A.; Hevia Méndez, M.L.; Míguez Santiyán, M.P. Agentes causales de intoxicaciones mortales en ganado bovino en España. *Inf. Tec. Econ. Agrar.* **2017**, *113*, 228–243. [CrossRef]
14. Google. The Keywords Used Were: "Cattle Sorghum Poisoning", "Intoxication Sorgho Bovins", "Intoxicação Sorgo Bovinos" and "Intoxicacíon Sorgo Bovino". Each Search Was Run Twice, with a Second One Adding "2023" in the String. Available online: www.google.com (accessed on 22 September 2023).
15. da Nóbrega, J.E., Jr.; Riet-Correa, F.; Medeiros, R.M.T.; Dantas, A.F.M. Intoxicação por *Sorhgum halepense* (Poaceae) em bovinos no semi-árido. *Pesqui. Veterinária Bras.* **2006**, *26*, 201–204. [CrossRef]
16. Juffo, G.D.; Pavarini, S.P.; Wouters, F.; Oliveira, L.G.S.; Antoniassi, N.A.B.; Cruz, C.E.F.; Driemeier, D. Intoxicação espontânea por *Sorghum sudanense* em bovinos leiteiros no Rio Grande do Sul. *Pesqui. Veterinária Bras.* **2012**, *32*, 217–220. [CrossRef]
17. Riet-Correa, F.; Machado, M.; Micheloud, J.F. Plants causing poisoning outbreaks of livestock in South America: A review. *Toxicon X* **2023**, *17*, 100150. [CrossRef] [PubMed]
18. Muwel, N.; Verma, R.; Nigwal, D.; Waskel, L.; Gurjar, R.; Ingole, M. Sorghum poisoning in buffaloes and its treatment. *J. Pharmacogn. Phytochem.* **2018**, *7*, 3737–3739.
19. Sinha, R.K.; Kumari, B.; Kumar, A.; Kumar, A.; Azad, C.S. Sorghum Poisoning in Cattle and its Therapeutic Management. *JAS* **2019**, *6*, 108–109.
20. Rajasokkappan, S.; Rajan, T.; Raghavendran, V.B. Sorghum poisoning in a cow and its successful management. *Pharma Innov.* **2020**, *9*, 164–165.
21. OECD (Suzy Cultivar). Available online: www.niab.com/oecdv2/variety/details/100558/suzy/0/0/vs/ALL/ALL/0 (accessed on 22 December 2023).
22. Giantin, S.; Franzin, A.; Topi, G.; Fedrizzi, G.; Nebbia, C. Outbreaks of lethal cyanogenic glycosides poisonings of cattle after ingestion of Sorghum ssp. Grown under drought conditions in August 2022 in Piedmont (North-Western Italy). *LAR* **2023**, *29*, 171–175.
23. ARPA Piemonte. Rapporto Siccità in Piemonte 2022—A Cura del Dipartimento Rischi Naturali e Ambientali, Torino 2022. Available online: www.arpa.piemonte.it/pubblicazioni-2/relazioni-tecniche/analisi-eventi/eventi-2022/rapporto-siccita-in-piemonte-2022 (accessed on 21 November 2023).
24. OECD (Piper Cultivar). Available online: www.niab.com/oecdv2/variety/details/2691/piper/1/1/vs/ALL/ALL/0 (accessed on 22 December 2023).
25. *SANTE/12089/2016*; Guidance Document on Identification of Mycotoxin in Food and Feed. European Commission. Health and Food Safety Directorate-General (DG SANTE): Bruxelles, Belgium, 2016.
26. European Union Reference Laboratory (EURL) for Mycotoxins and Plant Toxins. *EURL MP-Background Doc_003 Guidance Document Performance Criteria v1.1 Draft*; Wageningen Food and Safety Research: Wageningen, The Netherlands, 2021.
27. EFSA Panel on Contaminants in the Food Chain; Schrenk, D.; Bignami, M.; Bodin, L.; Chipman, J.K.; del Mazo, J.; Grasl-Kraupp, B.; Hogstrand, C.; Hoogenboom, L.R.; Leblanc, J.-C.; et al. Scientific opinion on the evaluation of the health risks related to the presence of cyanogenic glycosides in foods other than raw apricot kernels. *EFSA J.* **2019**, *17*, e05662. [PubMed]
28. Gensa, U. Review on Cyanide Poisoning in Ruminants. *J. Biol. Agric. Healthc.* **2019**, *9*, 1.
29. Vetter, J. Plant cyanogenic glycosides. *Toxicon* **2000**, *38*, 11–36. [CrossRef] [PubMed]
30. Hatch, R.G. Poisons causing respiratory insufficiency. In *Veterinary Pharmacology and Therapeutics*, 5th ed.; Booth, N.H., McDonald, L.E., Eds.; The Iowa State University Press: Ames, IA, USA, 1982; pp. 933–975.
31. Pickrell, J.A.; Oehme, F. Cyanogenic glycosides. In *Clinical Veterinary Toxicology*, 1st ed.; Plumlee, K.H., Ed.; Mosby: St. Louis, MI, USA, 2004; pp. 391–392.
32. Roder, J.D. *Veterinary Toxicology*, 1st ed.; Butterworth-Heinemann: Boston, MA, USA, 2001; pp. 138–142.
33. Nicholson, S.S. Southeastern Plants toxic to ruminants. *Vet. Clin. Food Anim.* **2011**, *27*, 447–458. [CrossRef] [PubMed]
34. European Parliament and Council. European Commission Directive N° 32/2002 on undesirable substances in animal feed. *Off. J. Eur. Union L* **2002**, *140*, 10–21.
35. Al-Sultan, S.I. *Sorghum halepensis* and its cyanide content. *Pak. J. Nutr.* **2003**, *2*, 123–124.
36. Zanandrea Sementi. Available online: https://zanandreasementi.it/prodotti/suzy-sorgo-foraggero-multi-sfalcio-da-fasciato-trinciato/ (accessed on 22 December 2023).
37. Peerzada, A.M.; Ali, H.H.; Hanif, Z.; Bajwa, A.A.; Kebaso, L.; Frimpong, D.; Iqbal, N.; Namubiru, H.; Hashim, S.; Rasool, G.; et al. Eco-biology, impact, and management of *Sorghum halepense* (L.) Pers. *Biol. Invasions* **2023**, *25*, 955–973. [CrossRef]

38. Jadav, C.N.; Makwana, R.B.; Parikh, S.S. Assessment of Hydrocyanic Acid (HCN) Content during Different Stages of Growth in Johnson Grass (*Sorghum halepensis*). *Agric. Rev.* **2022**, *43*, 394–397.
39. *EUR 31192 EN*; Toreti, A.; Bavera, D.; Acosta Navarro, J.; Cammalleri, C.; de Jager, A.; Di Ciollo, C. Drought in Europe August 2022. Publications Office of the European Union: Luxembourg, 2022.

animals

MDPI

Article

Retrospective Study of 25 Cases of Acorn Intoxication Colitis in Horses between 2011 and 2018 and Factors Associated with Non-Survival

Tanguy Hermange *, Basile Ruault and Anne Couroucé

Unité de Nutrition, PhysioPathologie et Pharmacologie, Ecole Nationale Vétérinaire, Agro-Alimentaire et de L'alimentation Nantes-Atlantique (Oniris), Université Bretagne Loire, 44307 Nantes, France; basileruault@yahoo.fr (B.R.); acourouce@orange.fr (A.C.)
* Correspondence: tanguy.hermange@gmail.com

Simple Summary: The aim of this study is to describe acorn intoxication colitis cases in horses and to find variables associated with non-survival. Data from horses presented at an equine hospital from 2011 to 2018 with a final diagnosis of acorn intoxication were included. Diagnosis was based on the following: season, the presence of acorns in the environment, clinical and hemato-biochemical parameters suggestive of a digestive/renal disease, the co-morbidity of companion animals, and post-mortem findings. A total of 25 horses were included. Results suggest that the intoxication may vary from year to year and that the number of cases seems to increase. Clinical signs associated with acorn intoxication were signs of circulatory shock, digestive signs, and abnormal temperature. Several significant clinical pathological findings were also described. Overall, 44% of horses survived. The majority of non-survivors died, or were euthanized, during the first 48 h. The following findings were significantly associated with non-survival: age, hemorrhagic diarrhea, heart rate, hematocrit, creatinine, blood lactate, and thickness of the colon wall at ultrasonography. This study provides equine practitioners with valuable prognostic information in cases of acorn intoxication.

Abstract: The aim of this study is to describe clinical data associated with acorn intoxication and to find variables associated with survival. Data from horses presented at CISCO-ONIRIS from 2011 to 2018 with a diagnosis of acorn intoxication were included. Diagnosis was based on the following: season, the presence of acorns in the environment, clinical and hemato-biochemical parameters suggestive of a digestive/renal disease, the co-morbidity of companion animals, and post-mortem findings. Statistical analysis was completed using Student's t-test for mean comparisons and a Chi-square test for group comparisons ($p < 0.05$). A total of 25 horses were included, and seasonality suggests that the intoxication may vary from year to year. Clinical signs associated with acorn intoxication were signs of circulatory shock (lethargy, tachycardia, abnormal mucous membrane, tachypnea), digestive signs (diarrhea, ileus, colic), and abnormal temperature. Clinical pathological findings included increased hematocrit, WBC, creatinine, BUN, GGT, AST, CK and decreased albumin. Overall, 44% (11/25) of horses survived. The majority (13/14) of non-survivors died, or were euthanized, during the first 48 h. Findings significantly associated with non-survival were age, heart rate, hemorrhagic diarrhea, ileus, hematocrit, creatinine, blood lactate, and thickness of the colon wall at ultrasonography. This study provides equine practitioners with valuable prognostic information in cases of acorn intoxication.

Keywords: acorn; colitis; equine; intoxication; prognosis; non-survival

1. Introduction

Acorn intoxication is a well-known concern for ruminants such as cattle and sheep [1], but information in horses is scarce. Cases [2–8] were reported a few decades ago with limited clinical information. More recently, two case series have been described. The first one is a case review [9] of nine intoxicated horses hospitalized in the UK, and the second one is a post-mortem case review [10] of 19 horses intoxicated and necropsied in France. However, the number of horses is limited in the first study, and clinical information and prognostic factors are not detailed in both studies.

Despite some hypotheses, the pathogenesis of acorn intoxication is not well understood. In other species, the toxicity has been attributed to its richness in gallotannins, especially digallic acid, which is hydrolyzed in the digestive tract, in pyrogallol with hemolytic properties and in gallic acid, which increased vascular permeability [11,12]. Tannins are macromolecules with strong bitterness and astringency properties which can also act with other proteins or enzymes, such as salivary proteins, endothelial cell proteins of the intestinal mucosa, or microbial proteins. Theses mechanisms could cause mucosal lesions, decrease digestion and transit, perturb gut microbiota, and induce renal tubular necrosis [11,13,14].

Toxicity varies between species and individuals. Clinical data are well described in ruminants [1] with gastroenteritis, constipation, and renal tubular necrosis, and other species such as pigs, which seem to be resistant to toxicity. This resistance has been attributed to tannin-binding salivary proteins (TBSPs) which can fix and hydrolyze tannins. Interestingly, an adaptation with inducible TBSPs when animals are exposed to a small account of acorns hulls has been shown in pigs and rodents [14,15]. TBSPs have not been investigated in horses.

Our retrospective study aims to describe the clinical data and to find prognostic factors associated with the non-survival of 25 horses suspected of acorn toxicity hospitalized in the CISCO (Oniris International Center for Equine Health) referral hospital between 2011 and 2018.

2. Materials and Methods

2.1. Data Collection

Medical records of the CISCO-ONIRIS referral hospital between 2011 and 2018 were reviewed for cases with a final diagnosis of acorn intoxication. Based on another study [9], criteria of inclusion were a final diagnosis of acorn intoxication with 4 of the 6 following criteria: illness during fall, presence of acorn hulls in the environment, co-mortality or co-morbidity of companion animals, acorn hulls in the feces or digestive tract, clinical and hemato-biochemical variables suggestive of a digestive and/or renal disease, and post-mortem findings suggestive of acorn intoxication.

All data obtained from medical records for history, clinical examination, blood analysis, complementary exams, outcome, and post-mortem information were reviewed. History data included seasonality, environment, time of admission, age, sex and breed. Clinical data included rectal temperature, mucous membrane aspect, capillary refill time (CRT), heart rate (HR), respiratory rate (RR), and digestive signs (diarrhea, colic). For statistical evaluation, mucous membrane abnormalities were graded with a score of 0 if normal, a score of 1 for mild abnormalities (dryness, paleness, or congestion alone) and a score of 2 for marked abnormalities (petechia/suffusion, combination of dryness/paleness/congestion, icterus, cyanosis).

Blood analysis included hematology, biochemistry panel (total protein, albumin, urea, creatinine, transaminase (ASAT), gamma glutamyl-transferase (GGT), alkaline phosphatase (ALP), creatine kinase (CK), fibrinogen, and electrolyte (sodium, potassium, chloride, calcium). Other complementary exams included transrectal palpation, naso-gastric intubation, transabdominal ultrasound, and paracentesis.

2.2. Statistical Analysis

Normality of continuous data distribution was evaluated using the Shapiro–Wilk W test. If necessary, data were log 10 transformed to normalize distribution. R® software (version 4.0.3) was used to perform the different statistical tests with Student's t-tests for mean comparisons and Chi-square tests for group (survivals versus non-survivals) comparisons. A p-value ≤ 0.05 was considered significant.

3. Results

3.1. History

Between 2011 and 2018, a total of 25 horses met at least four of the six inclusion criteria with the following distribution: illness during fall (n = 25/25), presence of acorn hulls in the environment (n = 25/25), co-mortality or co-morbidity of companion animals (n = 5/25), acorn hulls in the feces or digestive tract (n = 17/25), clinical and hemato-biochemical variables suggestive of a digestive and/or renal disease (n = 25/25), and post-mortem findings suggestive of acorn intoxication (n = 6/25).

All horses lived in pastures. A marked seasonality was noted with all horses presented between September 19th and October 29th. A large diversity of breeds was represented (seven French warmbloods, three Arabian horses, two Thoroughbreds, eight ponies, three Shetlands, one Welsh X Cob, and one Halfinger) typical of the hospital population. Males were overrepresented with 7 stallions, 10 geldings, and 8 females. However, the proportion of non-survivors in each group was similar (4/7 males, 6/10 geldings and 4/8 females died or euthanized).

The mean age was 16.0 ± 7 years old. Younger age was associated with survival; the survival group (n = 11) age was 12.4 ± 6.0 years old, while the non-survival groups (n = 14) were aged at 18.8 ± 6.6 years old (p = 0.01). Furthermore, beyond 13 years, there was a significantly ($p < 0.05$) higher mortality rate (n = 12 non-survivors of 16 horses beyond 13 years old versus 2/9 under 13 years old) with an odds ratio (OR) of 11.0 (Figure 1). The disease duration before admission varied between 4 and 144 h with a mean duration time of 6 h. Duration time of the disease before admission was not associated with survival whatever the period considered: >6 h (n = 12; p = 0.82); >12 h (n = 9; p = 0.33); >24 h (n = 8; p = 0.20).

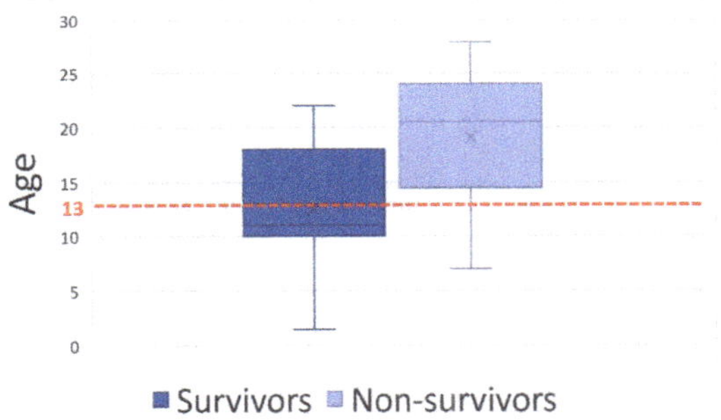

Figure 1. Age box-plot for age between survivors and non-survivors with the 13-year-old threshold in red dotted line.

3.2. Clinical Examination

Clinical signs of toxic shock were present with lethargy (25/25), tachycardia (20/23), abnormal mucous membrane (24/25), tachypnea (13/19), and abnormal rectal temperature (4/20). Mean HR at admission was 73 ± 19 bpm with a significant difference (p = 0.03) between survivors (63 ± 17 bpm) and non-survivors (80 ± 17 bpm). Considering an HR threshold of 65 bpm, there was a significant difference between groups (p = 0.01) with a positive predictive value for non-survival of 79% (Table 1) and an odds ratio of death of 12.8. Marked abnormalities of the mucous membrane (hemorrhagic border, congestion, icterus, cyanosis) versus mild abnormalities (dryness, paleness, or congestion alone) were associated with non-survival (p = 0.04). CRT rates higher than 3 s at admission were also associated with non-survival ($p < 0.05$). Two horses presented acute pulmonary edema and died quickly after admission (<3 h) despite intensive care. All the horses having abnormal rectal temperatures at admission (2 horses < 36.5 °C; 2 horses > 38.5 °C) died or were euthanized.

Table 1. Sensitivity, specificity, positive predictive value, negative predictive value of the prognostic factors identified with their respective threshold.

Parameters Thresholds	Sensitivity	Specificity	PPV	NPV	*p*-Value (Chi-Square Test)	Odds Ratios
HR > 65 bpm	85	70	79	57	0.01	12.8
Hematocrit > 60%	64	100	100	56	0.01	7.5
Lactate > 4.5 mmol/L	57	100	100	53	0.02	6.7
Creatinine > 229.8 µmol/L	57	91	89	56	0.05	5.5
Colon wall thickness > 22 mm	80	83	89	59	0.01	10.5

PPV: positive predictive value; NPV: negative predictive value; HR: heart rate; bpm: beats per minute.

Considering digestive signs, 68% (17/25) showed diarrhea, including 6 with hemorrhagic diarrhea, 28% (7/25) had signs of colic, and 28% (7/25) showed signs of ileus. All horses with hemorrhagic diarrhea died. For the seven horses which presented ileus at admission, five horses stayed in ileus until death and two horses developed diarrhea. Only one of the two horses who developed diarrhea survived. Ileus at admission was associated with non-survival ($p = 0.05$).

3.3. Blood Analysis

Hemato-biochemical analysis pointed to dehydration with a combination of increased hematocrit and blood lactate concentration, neutrophil leukocytosis, azotemia with increased urea and creatinine, hypoalbuminemia, and increased hepatic and muscular enzymes (Table 2). Electrolyte disturbances were present in several individuals, but only total calcium was decreased (2.33 ± 0.25 mmol/L). Marked intravascular hemolysis was reported in three horses.

Table 2. Hemato-biochemical results with mean value of total population, survivors and non-survivors.

	Mean Value ± SD	Survivors ± SD	Non-Survivors ± SD	*p*-Value (Student's *t*-Test)
Hematocrit (%)	52 ± 13	46 ± 9	57 ± 13	$p = 0.01$
Lactates (mmol/L)	4.7 ± 3.6	2.7 ± 1.5	6.4 ± 4	$p = 0.05$
Leucocytes (10^9/L)	13.7 ± 5.6	13.0 ± 5.5	14.4 ± 5.7	$p = 0.54$
Neutrophils (10^9/L)	9.6 ± 5.2	9.1 ± 5.6	9.9 ± 5.1	$p = 0.72$
Total proteins (g/L)	62 ± 9	61 ± 10	63 ± 8	$p = 0.55$
Albumin (g/L)	25 ± 5	24 ± 5	25 ± 5	$p = 0.60$
Creatinine (µmol/L)	211.3 ± 100.8	175.0 ± 87.5	239.6 ± 104.3	$p = 0.11$
Urea nitrogen (mmol/L)	22.5 ± 12.9	25.4 ± 13.6	$p = 0.22$	
GGT (µkat/L)	1.12 ± 1.12	1.29 ± 1.65	0.99 ± 0.55	$p = 0.54$
ASAT (µkat/L)	12.71 ± 29.81	19.56 ± 43.80	6.91 ± 4.91	$p = 0.31$
CK (µkat/L)	12.58 ± 9.72	9.59 ± 9.05	13.84 ± 10.10	$p = 0.32$

SD: standard deviation; GGT: gamma-glutamyl-transpeptidase; ASAT: aspartate amino-transferase; CK: creatine kinase.

Considering mean values, only hematocrit and blood lactate concentration were significantly different ($p = 0.01$) between survivors and non-survivors. When considering the following thresholds: hematocrit > 60%, blood lactate concentration > 4.5 mmol/L, and creatinine > 229.8 µmol/L, there was a significant difference ($p = 0.01$; $p = 0.02$; $p = 0.05$, respectively) between groups with lower chances of survival and respective odds ratios of 7.5, 6.7, and 5.5 (Figures 2–4). Using a threshold of hematocrit > 60%, blood lactate concentration > 4.5 mmol/L, and creatinine > 229.8 µmol/L, the positive predictive value for non-survival was above 90% for each variable (Table 1).

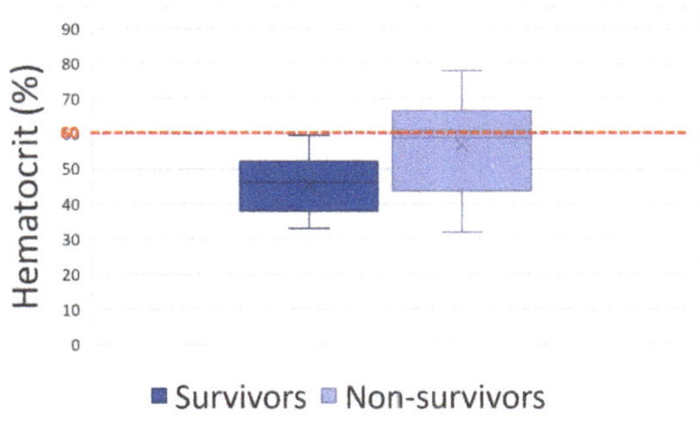

Figure 2. Hematocrit box-plot between survivors and non-survivors with the 60% hematocrit threshold in red dotted line.

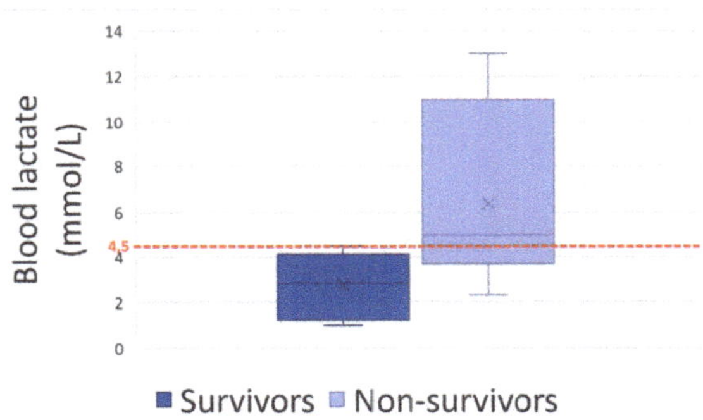

Figure 3. Blood lactates box-plot between survivors and non-survivors with the 4.5 mmol/L lactates threshold in red dotted line.

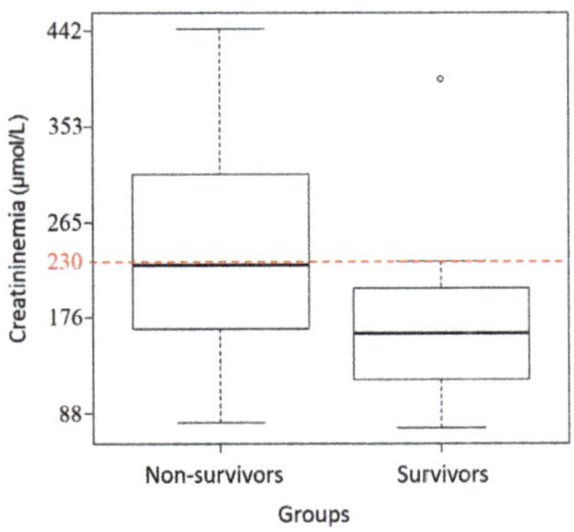

Figure 4. Creatinine box-plot between survivors and non-survivors with the 229.8 µmol/L (26 mg/L) creatinine threshold in red dotted line.

3.4. Others Complementary Exams

On rectal palpation, acorn hulls were present in 56% (14/25) of the cases. Thirty-five percent (5/17) of the horses had other abnormalities (three gas distention, one pelvic flexure impaction, one caecal impaction, and one tension band). The nasogastric tube did not yield any reflux for any (0/15) of the intoxicated horses, except for one when the horse was examined by its regular veterinarian before admission.

Thickening of the colon wall was significantly greater ($p = 0.02$) among the horses which did not survive (36 ± 19 mm versus 19 ± 6 mm) (Figure 5). Beyond 22 mm of thickness, the horses presenting such damages of the colon had significantly ($p = 0.01$) lower chances of survive with a positive predictive value of non-survival of 89% (Table 1). Thickening of the small intestine was observed for 35% (6/17) of the horses on which ultrasonography was performed. Abdominal ultrasonography showed severe thickening of the colon wall with values rarely encountered in other diseases with an average of 29 ± 17 mm (Figures 6 and 7).

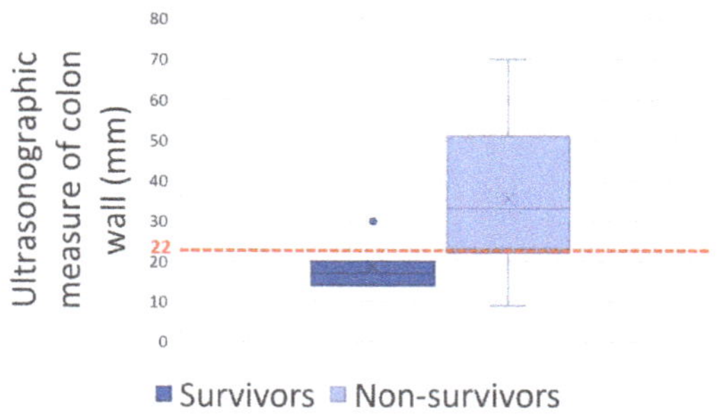

Figure 5. Colon wall thickening box-plot between survivors and non-survivors with the 22 mm thickening value threshold in red dotted line.

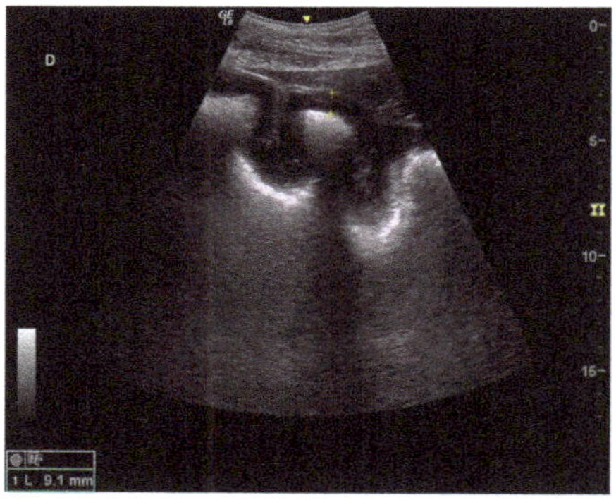

Figure 6. Ultrasound images of large colon. Moderate to marked thickening of large colon with easily visible villosities; the yellow dote line indicates a measurement of the colon wall at 9.1 mm.

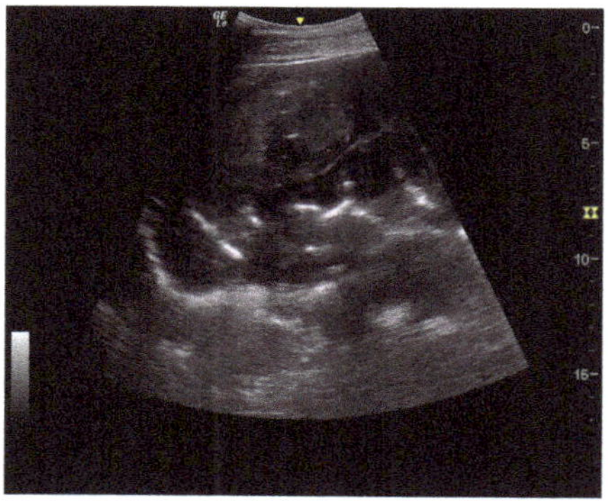

Figure 7. Ultrasound images of large colon. Very severe thickening of large colon.

Abdominal paracentesis was performed on four horses. Modified transudates were present in all horses, and peritoneal lactate concentration was measured. The two non-survivors had very high values (9.4 mmol/L and 10.7 mmol/L) whereas the two survivors had moderately increased values (3.2 mmol/L and 4.9 mmol/L). Unfortunately, the number of horses was not sufficient for statistical analysis.

3.5. Outcome

Fifty-six percent (14/25) of the horses died (n = 3) or were euthanized (n = 11), and the overall survival rate was 44%. The length of hospitalization varied from 1 h to 9 days, with a median duration of 48 h. Due to their critical conditions, three horses died or were euthanized in the first 3 h after admission, six others died during the first 24 h, and four more died between 24 and 48 h of hospitalization. Overall, 13/14 of non-survivors died or were euthanized during the first 48 h (Figure 8). Causes of euthanasia included pain non-responsive to treatment, persistence or deterioration of circulatory shock, severe complication such as esophageal lacerations (iatrogenic or secondary to ulcers) or iatrogenic rectal tears (grade 4/4), or respiratory difficulties due to severe pulmonary edema or pleural effusion (diagnosed ante-mortem). The only horse euthanized after 48 h was euthanized because of rectal laceration complications. Three quarters (8/11) of the horses who survived had normal parameters of hydration, proteinemia, or albuminemia after 48 h of hospitalization. Similarly, rapid return of appetite and production of feces during hospitalization seem to be good prognostic factors.

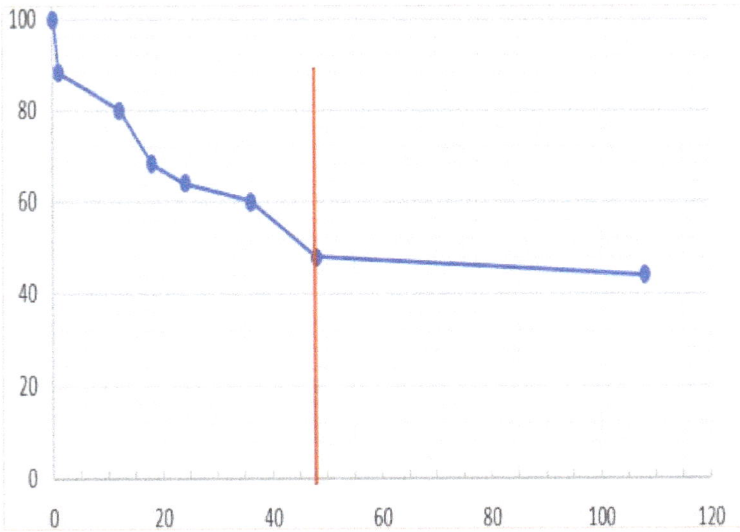

Figure 8. Proportion of horses still alive relative to duration of hospitalization (in hours); the red line marks the threshold of 48 h of hospitalization stay.

3.6. Post-Mortem

Post-mortem findings were available for five horses. A large amount of acorn husks (Figure 9) were observed in the intestinal tract of all horses (n = 5/5). Main lesions observed were an extensive severe edematous colitis in four of the five horses (Figures 10–12). The cadaver of the last one was too putrefied for post-mortem examination. Extensive ulcers were present in two horses, while extensive hemorrhages appeared in two others. Other significant lesions included petechia, peritoneal or pleural effusion, renal congestion and adrenal hemorrhage.

Figure 9. Ascending colon content with numerous acorn hulls.

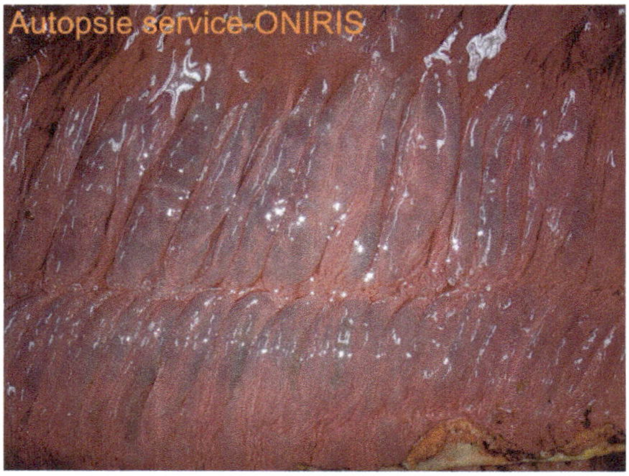

Figure 10. Picture of luminal aspect of ascending colon.

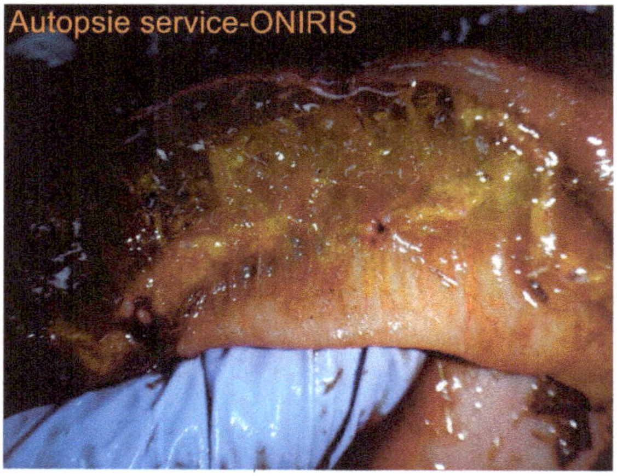

Figure 11. Transversal section of ascending colon.

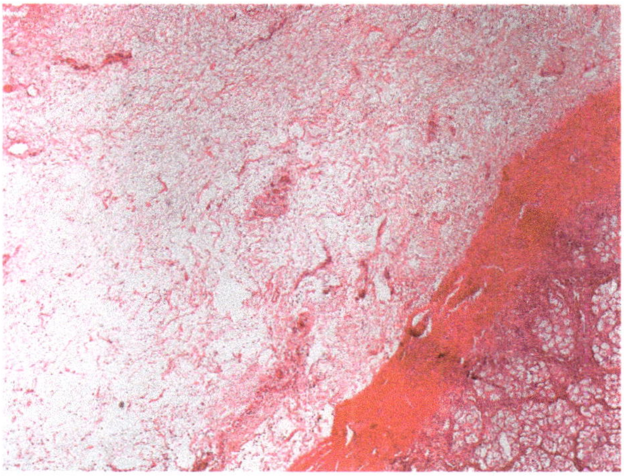

Figure 12. Histologic section of a transversal section; note the severe submucosal edema present in Figures 11 and 12.

4. Discussion

This study constitutes a relatively large clinical case review about acorn intoxication in equids and describes prognostic factors to help the equine practitioner assess disease severity. Ideally, definitive diagnosis would have required measurements of tannin metabolites in serum or urine samples. A gas chromatography/mass spectrometry test has been validated for this purpose in cattle [12] and has been used in one horse [2]. However, the test has not been validated for horses and is not routinely available in France. For these reasons and due to the retrospective nature of this study, analyses were not performed. Therefore, several inclusion criteria were chosen based on previous studies [2,9]. All horses met the three main criteria: illness during fall, presence of acorn hulls in the environment, and clinical and hemato-biochemical variables suggestive of a digestive and/or renal disease. The other criteria (presence of acorn in digestive tract, co-morbidity, and post-mortem findings) were not present in all horses. Due to a presumptive diagnosis and cost reasons, investigation of enteropathogens was not performed, either, but it should have been performed to exclude potential bias.

The annual variations observed and the increasing number of cases over the period from 2011 to 2018 are remarkable. The annual variations of intoxication may be due to several factors: acorn quantity, the presence of immature acorns, or acorn tannin concentrations. Data presented to the 2019 Havemeyer Workshop on acute colitis revealed that 78% (15/19) of owners reported that there was an unusual high quantity of acorns hulls when the horses became sick [16]. Sixty-three percent (12/19) thought they looked unusually green/immature, suggesting a role of the quantity and of the aspect of the acorn hulls [16]. The 'mast effect' is a well-known phenomenon for botanists. It characterizes the variations of acorn production depending on years with high production every 2 to 5 years depending on oak species and meteorological conditions [17–21]. Moreover, it has been shown that tannins and phenolic acid concentrations vary between oak species [22]. Interestingly, some owners reported that some intoxicated horses lived in their pasture for years with an access and consumption of acorn hulls some years. These horses only developed intoxication recently, suggesting that tannins type or concentrations may vary over time. Also, the increasing number of cases over time in this study, with the two others recent case reviews [19,20], may suggest a new emerging form of acorn intoxication. Although underreported cases may also remain a hypothesis, the absence of cases in the database of our hospital and necropsy unit for decades make it unlikely.

From a clinical point of view, our study did not reveal any type, breed or sex predisposition. A marked seasonality was observed with all cases admitted from September 19 to October 29, which may help the practitioner pay attention to acorn intoxication during this period in order to prevent and/or diagnose the disease. The mean age was 16.0 ± 7 years old, which concurs with the two other reviews [9,10]. Older age correlates with non-survival with an odds ratio of 11.0 beyond 13 years old. Case analysis revealed that the critical clinical status was the cause of euthanasia in all horses, excluding a potential financial bias. Reasons for this older age susceptibility remain unexplained. In human septic shock, older age has been shown to be a negative prognostic factor, and poorer physiological adaptation to circulatory shock is suspected [23]. To our knowledge, a similar phenomenon has not been investigated in horses.

In addition to the marked seasonality, the clinical presentation was quite typical with an acute to fulminant severe edematous colitis leading to toxic shock, which was complicated in several cases with marked azotemia, increase hepatic or muscle enzymes, and intravascular hemolysis. Although not fully detailed in other studies, the hemato-biochemical results were similar to previous reports [2,9] and can be easily understood in view of the acute colitis shock. Abdominal ultrasonography examination revealed marked thickening of the large colon wall, especially in non-survivors with values rarely encountered in other diseases. The thickness was so severe that clear measurement was sometimes difficult (Figure 7). This reflects the severe colon edema as it could be observed post-mortem (Figures 11 and 12).

Older age, signs of shock (HR, hematocrit, blood lactates, blood creatinine), hemorrhagic diarrhea, ileus, and increased colon wall thickness at the transabdominal ultrasonography exam were associated with non-survival and may help practitioners assess prognosis. Using a threshold of hematocrit > 60%, blood lactate > 4.5 mmol/L, creatinine > 229.8 μmol/L, and colon wall thickness > 22 mm places the positive predictive value for non-survival above 89% for each variable.

The overall survival rate was low at 44% but a little superior to the 33% survival rate recently described [9]. The first 48 h appear very critical with the large majority of horses (13/14) dying during this period. As seemingly presented in a previous report [9], the improvement of clinical variables during this period may also be considered as a good prognostic factor.

The main limitation of this study includes the relative low number of horses for statistical analysis which precluded multivariable analyses. However, it remains the larger study of acorn intoxication colitis, and several factors were found to be statistically significant in univariate analyses. Other limitations of this study are inherent to its retrospective and clinical aspects, with inter-individual variability regarding the cases' assessment and management. Indeed, due to the emergency aspect of the intoxication, the first clinical exam and assessment of horses were not performed by the same clinician every time. However, after emergency hours, all the horses were managed by three experienced practitioners, thereby reducing this potential bias. Other limitations include the lack of a definitive diagnosis with tannin metabolites search and the lack of the exclusion of enteropathogens etiology. However, in our practitioner opinion, the complete clinical description of the patients and the focus on four of six inclusion criteria provide a very likely diagnosis of acorn toxicity.

5. Conclusions

This study presents a unique large clinical case review of acorn intoxication in horses. It provides valuable clinical data and prognostic information, which could help equine practitioners in diagnostic procedure and prognostic assessment. The intoxication is particularly severe. Despite intensive care, a large proportion of horses died or were euthanized, and findings associated with non-survival were age, heart rate, hemorrhagic diarrhea, ileus, hematocrit, creatinine, blood lactate, and increased colon wall thickness on ultrasonography.

Interestingly, the intoxication presents an apparent increased number of cases over years and an annual variation which may be related to the 'mast effect'. The study of tannins metabolites in horse's serum and in acorn hulls—ideally during years with and without intoxication cases—is now necessary to have a better understanding of the epidemiology and physiopathology of the disease.

Author Contributions: Conceptualization, T.H., B.R. and A.C.; methodology, B.R.; software, B.R.; validation, T.H., B.R. and A.C.; formal analysis, B.R.; investigation, B.R.; writing—original draft preparation, T.H. and B.R.; writing—review and editing, T.H.; supervision, A.C. All authors have read and agreed to the published version of the manuscript.

Funding: This research received no external funding.

Institutional Review Board Statement: Not applicable.

Informed Consent Statement: Not applicable.

Data Availability Statement: The data presented in this study are available on request from the corresponding author (accurately indicate status).

Acknowledgments: We thank all of the CISCO hospital team, Autopsie Service, and the owners for their trust.

Conflicts of Interest: The authors declared no potential conflicts of interest with respect to the research, authorship, and/or publication of this article.

References

1. Spier, S.J.; Smith, B.P.; Seawright, A.A. Oak toxicosis in cattle in northern California: Clinical and pathologic findings. *J. Am. Vet. Med. Ass.* **1987**, *191*, 958–964.
2. Anderson, G.A.; Mount, M.E.; Vrins, A.A. Fatal acorn poisoning in a horse: Pathologic findings and diagnostic considerations. *J. Am. Vet. Med. Ass.* **1983**, *182*, 1105–1110.
3. Broughton, J. Acorn poisoning. *Vet. Rec.* **1976**, *99*, 403–404. [CrossRef] [PubMed]
4. Daniels, M.G. Acorn poisoning. *Vet. Rec.* **1976**, *99*, 465–466. [CrossRef] [PubMed]
5. Duncan, C.S. Oak leaf poisoning in two horses. *Cornell Vet.* **1961**, *51*, 159–162. [PubMed]
6. Warren, C.G.; Vaughan, S.M. Acorn poisoning. *Vet. Rec.* **1985**, *116*, 82. [CrossRef] [PubMed]
7. Wharmby, M. Acorn poisoning. *Vet. Rec.* **1976**, *99*, 343. [CrossRef]
8. Wiseman, A.; Thompson, H. Acorn poisoning. *Vet. Rec.* **1984**, *115*, 605. [CrossRef]
9. Smith, S.; Naylor, R.J.; Knowles, E.J.; Mair, T.S.; Cahalan, S.D.; Fews, D.; Dunkel, B. Suspected acorn toxicity in nine horses. *Equine Vet. J.* **2015**, *47*, 568–572. [CrossRef]
10. Linster, M.; Foucher, N.; Le Net, J.-L.; Tapprest, J. Intoxication par les glands chez les équidés: Étude à partir de 19 cas autopsiés entre 2007 et 2017 (Acorn toxicity in horses: Study of 19 cases autopsied between 2007 and 2017). *Le Nouv. Pract. Vétérinaire Équine* **2018**, *12*, 10. (In French)
11. McSweeney, C.S.; Palmer, B.; McNeill, D.M.; Krause, D.O. Microbial interactions with tannins: Nutritional consequences for ruminants. *Anim. Feed Sci. Technol.* **2001**, *91*, 83–93. [CrossRef]
12. Tor, E.R.; Francis, T.M.; Holstege, D.M.; Galey, F.D. GC/MS determination of pyrogallol gallic acid in biological matrices as diagnostic indicators of oak exposure. *J. Agric. Food Chem.* **1996**, *44*, 1275–1279. [CrossRef]
13. Cappai, M.G.; Wolf, P.; Liesner, V.G.; Kastner, A.; Nieddu, G.; Pinna, W.; Kamphues, J. Effect of whole acorns (*Quercus pubescens*) shred based diet on parotid gland in growing pigs in relation to tannins. *Livest. Sci.* **2010**, *134*, 183–186. [CrossRef]
14. Shimada, T. Salivary proteins as a defense against dietary tannins. *J. Chem. Ecol.* **2006**, *32*, 1149–1163. [CrossRef] [PubMed]
15. Mehansho, H.; Asquith, T.N.; Butler, L.G.; Rogler, J.C.; Carlson, D.M. Tannin-mediated induction of proline-rich protein-synthesis. *J. Agric. Food Chem.* **1992**, *40*, 93–97. [CrossRef]
16. Hermange, T.; Ruault, B.; Flambard, P.; Courouce, A. Epidemiological analysis of 25 cases of acorn intoxication colitis presented at CISCO-ONIRIS in France between 2011 and 2018. Proceeding of the 2019 Havemeyer workshop on acute equine colitis, Niagara-on-the-lake, ON, Canada, 14–17 July 2019; Wiley Online Library: New York, NY, USA, 2019.
17. Espalta, J.M.; Cortès, P.; Molowny-Horas, R.; Sanchez-Humanes, B.; Retana, J. Masting mediated by summer drought reduces acorn predation in mediterranean oak forests. *Ecology* **2008**, *89*, 805–817. [CrossRef]
18. Koenig, W.D.; Knops, J.M.H.; Carmen, W.J.; Stanback, T.; Mumme, R.L. Acorn production by oaks in central coastal California: Influence of weather at three levels. *Can. J. For.* **1996**, *26*, 1677–1683. [CrossRef]
19. Koenig, W.D.; Knops, J.M.H. Large-scale spatial synchrony and cross-synchrony in acorn production by two California oaks. *Ecology* **2013**, *94*, 83–93. [CrossRef]

20. Koenig, W.D.; Knops, J.M.H.; Carmen, W.J.; Pearse, I.S. What drives masting? The phenological synchrony hypothesis. *Ecology* **2015**, *96*, 184–192. [CrossRef]
21. Perez-Ramos, I.M.; Padilla-Diaz, C.M.; Koenig, W.D. Environmental drivers of mast-seeding in Mediterranean oak species Does leaf habit matter? *J. Ecol.* **2015**, *103*, 691–700. [CrossRef]
22. Basden, K.W.; Dalvi, R.R. Dermination of total phenolics in acorns from different species in conunction with acron poisoning in cattle. *Vet. Hum. Toxicol.* **1987**, *29*, 305–306.
23. Pavon, A.; Binquet, C.; Quenot, J.-P. Épidémiologie du choc septique: Facteurs pronostiques à trois mois. (Epidemiology of septic choc: 3-months prognostic factors). *Réanimation* **2012**, *22*, S17–S20. (In French)

animals

MDPI

Review

Do Poisonous Plants in Pastures Communicate Their Toxicity? Meta-Study and Evaluation of Poisoning Cases in Central Europe

Sabine Aboling

Institute of Animal Nutrition, University of Veterinary Medicine Hannover, Bischofsholer Damm 15, 30173 Hannover, Germany; sabine.aboling@tiho-hannover.de

Simple Summary: Secondary plant metabolites, which can be toxins, may exert a repellent function. However, lethal cases of pasture poisoning show that this protection may fail. The reasons behind this are unknown. The aim of this review was (1) to document under which circumstances poisoning occurred in cattle, sheep, goats, and horses and (2) to present a checklist of poisonous pasture plant species that may occur in Central Europe. The checklist comprised 52 taxa, of which 13 taxa were associated with no poisoning owing to the lack of a report, 11 taxa with evidence-based zero poisoning by avoidance or ingestion, and 28 taxa with poisoning due to ingestion. Nine plant taxa caused poisoning in pastures in more than 100 individuals, for example, sycamore (*Acer pseudoplatanus*) in horses, cowbane (*Cicuta virosa*) in cattle, and St. John's wort (*Hypericum perforatum*) in sheep. Zero poisoning accounted for 40%, and poisoning accounted for 60% of all 85 cases. Poisoning was most often associated with a limited choice of feed (24.7%) and least often associated with the co-ingestion of grass (4.7%). In between were the circumstances of "overgrazing" (12.9%) and "seasonally scarce feed" (10.6%). The results of this inquiry suggest that poisoning virtually always evolves owing to hunger, while 40% zero poisoning proved successful co-existence to a large extent. Therefore, poisonous plants in pastures do communicate their toxicity under animal-friendly grazing management. With the help of the checklist, farmers may evaluate the risk posed by poisonous plants to their animals.

Abstract: One of the possible roles of secondary plant metabolites, including toxins, is facilitating plant–animal communication. Lethal cases of pasture poisoning show that the message is not always successfully conveyed. As the focus of poisoning lies in the clinical aspects, the external circumstances of pasture poisoning are widely unknown. To document poisoning conditions in cattle, sheep, goats, and horses on pastures and to compile a checklist of plants involved in either poisoning or co-existence (zero poisoning), published case reports were evaluated as primary sources. The number of affected animal individuals was estimated within abundance classes from 0 to more than 100. The checklist of poisonous plants comprised 52 taxa. Of these, 13 taxa were deemed safe (no reference was found indicating poisoning), 11 taxa were associated with evidence-based zero poisoning (positive list), and 28 taxa were associated with poisoning (negative list). Nine plant taxa caused poisoning in more than 100 animal individuals. Zero poisoning accounted for 40% and poisoning accounted for 60% of a total of 85 cases. Poisoning was most often associated with a limited choice of feed (24.7%), followed by overgrazing (12.9%), seasonally scarce feed (10.6%), and co-ingestion of grass (4.7%). Hunger interferes with plant–animal co-existence, while zero poisoning improves it. In conclusion, poisonous plants in pastures may communicate their toxicity if the animals have enough alternative feed plants. An individual animal might utterly perceive the communication of toxicity by the plant species but be forced to ignore the message owing to a limited choice of feed options.

Keywords: co-existence; overgrazing; zero poisoning; co-ingestion; case report; checklist; plant-animal communication

1. Introduction

Relying on pre- and post-ingestive sensations, herbivorous vertebrates thrive because of their capability to navigate between distinct kinds of plant species within vegetation [1]. This is enabled by the principle that certain volatile and non-volatile molecules act as "secondary plant metabolites, eliciting pharmacological or toxicological effects in humans and animals" [1]. Plant-animal communication is based on these metabolites, on the one hand, and on the senses of the animal, on the other hand. Certain poisonous metabolites that are thought to be species-specific have evolved to permit the co-existence of both certain animals and plants. As the sheer number of studies alone shows, the ecological role of those metabolites as repellents is still almost exclusively demonstrated in the case of non-vertebrates. In herbivorous vertebrates, one assumes a pre-ingestive sensation towards volatile, mostly chemically unidentified metabolites, so-called "attractants or repellents", leading to a distinct feeding decision [2]. However, in the case of some prominent poisonous plant species, the poison (e.g., oxalic acid in dock (*Rumex*), alkaloids in yew (*Taxus*) and ragwort (*Senecio*) and hypoglycin A in sycamore (*Acer pseudoplatanus*)) neither is volatile nor does it act as a repellent. On the other hand, we know the chemical and functional nature of some potential volatile repellents, such as monoterpenes, which pre-ingestively lead to feed aversions in both naïve red deer calves [3] and lambs [4].

In a post-ingestive sensation, herbivorous vertebrates sense metabolites with their taste organs [5,6]. Among relevant substances that would cause a positive so-called "post-ingestive feedback" [7] are nutrients [8], which act both via taste in the short term [8] and via satiety in the long term [7]. In contrast, some plant secondary metabolites may impose a negative post-ingestive feedback associated either with a bad taste or with illness or pain [7]. For example, drimane sesquiterpenes deter fishes from feeding on sponges and "may be a result of direct action on taste receptors" [9]. Tannins, as non-volatile compounds, are produced by acacias, leading to reduced browsing by giraffes [10]. Pre- and post-ingestive sensations are co-ordinated: deer smell the feed before tasting it [11]. Goats browsing on different species of juniper (*Juniperus*), distinguished by species-specific oil concentrations, did not show whether they distinguished different juniper species through odor, taste, or both [12].

In this respect, poisonous plants' nutrients may attract a grazing animal, while the plants' poisons may prevent severe or lethal poisoning by causing adaptive grazing behaviors ranging from complete avoidance to the ingestion of small quantities. The consequences for the animal during adaptation or learning are either invisible as subclinical reactions (no poisoning) or visible as non-specific symptoms, such as diarrhea or salivation (poisoning). Moreover, there are specific symptoms after the ingestion of certain plant species, such as brown urine in the case of pasture myopathy in horses through the ingestion of juvenile *Acer pseudoplatanus*, vomiting in the case of *Rhododendron* poisoning in small ruminants, and colic in horses through the ingestion of *Colchicum autumnale*.

Unfortunately, lethal cases of pasture poisoning show that protection by secondary plant metabolites may fail. Corresponding symptoms and diseases in the animal are described in detail, while other variables in the context are scarcely described, such as the biology of poisonous plants as well as the composition and state of the vegetation. This approach has the following three consequences:

(1) Plants are regarded toxicologically. Publications on poisonous plants rely on experimentally proven toxicities mostly in laboratory animals. This approach is termed "intoxication" and describes the artificial process of the targeted and forceful application by humans of a toxin to an animal. Moreover, feeding experiments serve to define the risk based on a dose-effect relationship. This leads to entries, such as "[w]ild plants that can cause poisoning in farm animals" [13], or to conclusions, such as "[b]ased on toxicological data, poisoning of livestock and companion animals by plants is a relatively common occurrence in Europe" [14]. Even though a study acknowledges the "[r]esponse of herbivores" to plant poisons, it lists records of "[t]oxic plants in Central Europe" [15].

(2) Case details are neglected. In a recent (2015) publication, poisoning with autumn crocus (*Colchicum autumnale*) is described as follows: "[p]oisoning may occur when the young spring leaves or autumn flowers are ingested in pastures [. . .] (3)" [16]. The indicated reference (3) above was published in 2012 and states "[p]oisoning of animals in the spring involves ingestion of the young leaves, whereas in the autumn the flowers of plants growing wild in pastures are implicated (Humphreys, 1988)" [17]. Humphreys [18] finally indicates the primary sources [19] and [20], stating that "*C. autumnale* poisoning in cattle is dealt with by Debarnot [20] and in sheep by Tribunskii [19]", but does not mention any details. These references from 1968 and 1970 revealed that the "animals" were lambs and cattle and that one single cow ate "flower after flower" on her pasture [20]. None of the references mentioned "young spring leaves".

(3) Context of poisoning is not recorded. A review from 2010 on "animal poisoning in Europe" listed "[. . .] one case report of sheep poisoned [. . .] by monks hood (*Aconitum napellus*)" [21]. The indicated source stated "[u]nfortunately not specific information could be collected at that time" [22]. Already in 2008, both those "notorious discrepancies of thinly documented information" were addressed, and the monography "International Poisonous Plants Checklist: An Evidence-Based Reference" was published along with original literature references, covering mainly cases of conserved feed [23]. In 2015, the first approach appeared, entitled "Exposure assessment of cattle via roughages to plants producing compounds of concern", that explicitly considered not only toxins as a variable in pasture poisoning but also ecologically relevant aspects such as the abundance of plant species in grasslands and the selection behavior of animals within trough-fed roughage [24].

These issues (toxicological focus, neglected case details, missing context) make it difficult to be aware of the outer circumstances under which the species-specific relationship between plants and herbivorous animals fails such as the kind and quality of feed plant species available on the pasture, the composition of the vegetation, or the presence of certain developmental stages of poisonous and feed plants. Therefore, animal keepers would emphasize rather incalculable dangers than the mutual benefits of grazing for both husbandry animals [25] and agricultural landscapes [26]. To evaluate the risks for animals on pastures, evidence-based data from everyday practice need to be compiled. Among the references, scientific case reports enjoy priority since they enable an inductive approach with authentic first-hand data. The aim of this review is (1) to document under which circumstances poisoning occurred in cattle, sheep, goats, and horses and (2) to present a checklist of poisonous pasture plant species in Central Europe from the point of view of these animals.

2. Materials and Methods

The pastures considered in this study were mainly located in Central Europe, geographically including, according to "Flora of Central Europe" [27], the countries Austria, Croatia, the Czech Republic, Germany, Hungary, Luxemburg, Poland, Slovakia, and Switzerland. Moreover, this study evaluated a few cases observed in the United Kingdom, in Norway and Sweden, and in Spain and Italy. Here, "pasture" is understood as any kind of grassland such as continuously growing herbaceous vegetation within a fenced area that is regularly grazed by large herbivores, no matter whether or not the site is mown. Pure meadows and year-round wild pastures are excluded due to their deviating floristic composition.

Scientific databases of references were browsed for plants, using the terms "poisonous" and "toxic" as well as "cattle", "sheep", "goats", and "horses" as search algorithms. Recorded plant taxa for the checklist needed to meet three criteria: (1) presence of secondary plant metabolites such as toxins that are potentially poisonous for farm animals, (2) growing in pastures, and (3) occurring in Central Europe. In the case of chemically similar species of one genus or family, only one or two genera appeared in the checklist.

In the case of chemically similar and closely related genera, both genera were listed as one entry.

Apart from primary sources such as case reports, papers, or brief communications on pasture poisoning, ecological or agricultural studies on grazing farm animals contributed to the checklist, too. In two of three cases, personal communications were included and labeled as such. In order to reproduce the authentic information from the references as authentically as possible, the results consisted largely of citations.

"Poisoning" comprised all surviving or lethal cases of pastured animals with reported clinical symptoms of poisoning due to a known ingestion of poisonous plants. While the names of the plant secondary metabolites are listed (Table S1), the corresponding symptoms of poisoning are not systematically described but are often part of the citations. Two kinds of references were regarded as proof of poisoning: (1) small formats, such as a correspondence as documented evidence, and (2) peer-reviewed papers as rationale, which provided evidence. However, it was not always possible to deduce whether certain sources had been peer-reviewed or not, particularly those published before 1950.

"Zero poisoning" (the absence of poisoning) comprised all cases of pastured animals with direct contact with poisonous plants but without known (reported or documented) symptoms of plant poisoning. Zero poisoning was either indirectly assumed due to the lack of a case report or directly proven on the basis of any publication as defined below (Table 1). If proven via a publication, zero poisoning was based either on the documented avoidance or on the documented inconsequential ingestion of a poisonous plant by an animal. If the authors did not reasonably report whether the animal had avoided or ingested the plant, an unknown reason for zero poisoning was assumed. In any case, published evidence-based incidents of zero poisoning were scrutinized to see whether the authors had indeed failed to mention any clinical or subclinical symptoms.

Table 1. Criteria for data selection and corr sponding number of individuals.

Label	Criteria of Data Selection and Definitions of Poisoning and Zero Poisoning	Animal Individuals (n)
A	0 cases assumed or yet to be documented if no publication reference could be found that mentioned the presence of both the plant species and the animal species on a pasture (indirect evidence): "assumed zero poisoning". In other words, from the lack of such a publication reference, the lack of a poisoning incidence has been assumed.	0
B	0 cases proven or yet to be documented ("proven zero poisoning" or synonymous "evidence-based zero poisoning") if a publication reference on the presence of both plant and animal species on a pasture was detected (direct evidence), with the proof of (1) avoidance of the plant, (2) ingestion of the plant, or (3) of any other kind of co-existence leaving unanswered whether the plant was avoided or ingested.	0
C	1–9 individuals in proven cases of pasture poisoning, mentioned in one or more publications.	<10
D	10–99 individuals in proven cases of pasture poisoning, mentioned in one or more publications.	>10
E	100-more individuals in proven cases of pasture poisoning, mentioned in one or more publications.	>100

Labels A–E are added for a better orientation.

The number of affected animal individuals in the checklist was estimated within abundance classes from 0 up to more than 100, cumulatively adding records documented in as many papers as could be found.

An example for criterion B "0 cases proven" is the following: "[s]uch animals [goats] are usually kept on marginal land where bracken [*Pteridium aquilinium*] can occur. As there seem to be no reports about the effect of bracken on goats and the extent, if any, to which bracken carcinogens may be transferred to the milk, such studies would be very valuable" [28]. The corresponding combination "goat and *Pteridium*" was recorded in this meta-study with both the reference [28] and the circumstance "unknown".

If both "zero poisoning" and "poisoning" had been detected in a certain plant-animal combination because of two distinct reports in two papers, the checklist would categorize the corresponding plant species as "poisonous" due to priority. However, the finding of "zero poisoning" with those plant species under different circumstances was supplemented (see Section 3.2.8).

Pigs as dominating important husbandry animals were omitted because only four studies evaluating plant-pig interactions on pastures were available ([29–32]).

Each identified plant species was searched for a relation to cattle, sheep, goats, and horses in order to check whether the combination led to a result. Grazing always had to be the dominant kind of feeding, but data on pasture animals stabled only during the winter served likewise as material. Cases of pasture poisoning counted as proven if the authors provided a substantial report, not merely anecdotal notes. Thus, statements such as "[p]oisoning due to eating roots of water hemlock is discussed" [30] were dismissed since it was unclear whether the cases they referred to had been pasture poisoning or not. Moreover, several field observations were revealed to not be cases of pasture poisoning and thus were not considered. For example, "P. described the illness of a bullock which died within three min. of eating water hemlock roots [*Cicuta virosa*] contained in mud from a recently cleaned out drinking place [. . .]" [31].

In particular, papers that provided inconclusive evidence were excluded. For example, a case report on hound's tongue (*Cynoglossum officinale*) reads (translated from German) as follows: "In a group of four horses, one horse showed severe flatulence along with symptoms of colic after some days of being pastured. Hound's tongue was not identifiable at the time of the site's inspection [three weeks later after the incidence] neither as rosettes nor as flowering shoots" [33]. Instead, the plant species had been identified only in the feed sample provided to the authors before the inspection of the site took place. Due to the contradictory findings, it remains open whether hound's tongue caused the symptoms or not.

Taxonomy of botanical scientific names followed "The International Plant Names Index and World Checklist of Vascular Plants 2023" [34]. The genus *Senecio* has been recently partly renamed in *Jacobaea* [34]. However, in the case of *Senecio alpinus* (now *Jacobaea alpinus* (L.) Moench.) and *Senecio jacobaea* (now *Jacobaea vulgaris* Gaertn.), these synonyms were still used here due to their widespread usage. The dictionary of plant names by Zander (2002) [35] provided the common names.

3. Results

The checklist of poisonous plants comprised 52 taxa (see Table S1). Related to four animal species, respectively, this resulted in 208 combinations (52 × 4), denoted as "cases" (Table 2).

Table 2. Animal individuals and cases of plant poisoning per animal species.

Label	Animal Individuals (n)	Cases (n)	Cases (%)	Cattle Cases (%)	Sheep Cases (%)	Goat Cases (%)	Horse Cases (%)
A	0 (assumed)	135	64.9	50.0	65.4	80.8	63.5
B	0 (proven)	34	16.3	15.4	17.31	15.4	17.3
C	<10	20	9.6	15.4	7.7	3.8	11.5
D	>10	8	3.8	9.6	0.0	0.0	5.8
E	>100	11	5.3	9.6	9.6	0.0	1.9
A–E	Total	208	100	100	100	100	100

Table 2. *Cont.*

Label	Animal Individuals (n)	Cases (n)	Cases (%)	Cattle Cases (%)	Sheep Cases (%)	Goat Cases (%)	Horse Cases (%)
A–B	0 (assumed + proven)	169	81.3	65.4	82.7	96.2	80.8
B–E	0 > 100 (proven)	73	35.1	50.0	34.6	19.2	36.5
C–E	1 > 100 (proven)	39	18.8	34.6	17.3	3.8	19.2

The letters A and B denote two kinds of evidence. The letters C-E denote distinct numbers of animal individuals. The labels are used to show at a glance the content of the three formed groups A–B, B–E, and C–E.

A total of 169 of these combinations (81.3%) represented assumed or proven zero poisoning (A–B). However, for most of the combinations (64.9%), no publication could be found. These cases counted as assumed zero poisoning (A). In contrast, a further 34 cases of zero poisoning were documented by at least one publication and counted as proven zero poisoning (B). Moreover, 39 cases of plant poisoning existed (C–E). Among the references that gave evidence in all 73 cases (B–E) were three personal communications from the author [36–38].

3.1. Checklist of Poisonous Plant Species in Pastures in Central Europe

Among the 52 taxa of the checklist, 13 taxa were associated with assumed zero poisoning: field maple (*Acer campestre*), mugwort (*Artemisia vulgaris*), Barbara's herb (*Barbarea*), hoary alison (*Berteroa incana*), adderwort (*Bistorta officinalis*), cuckoo flower (*Cardamine pratensis*), common horsetail (*Equisetum arvense*), spurge (*Euphorbia*), ground ivy (*Glechoma hederacea*), bristle grass (*Setaria*), comfrey (*Symphytum*), clover (*Trifolium*), and vetch (*Vicia*).

The remaining 39 plant taxa were involved in the above-mentioned total number of 73 evidence-based cases of both zero poisoning (11 taxa) and poisoning (28 taxa). In detail, the following taxa caused no poisoning: marsh marigold (*Caltha palustris*), giant hogweed (*Heracleum mantegazzianum*), horseshoe vetch (*Hippocrepis comosa*), poppy (*Papaver*), lousewort (*Pedicularis palustre*), smartweed/knotgrass (*Persicaria/Polygonum*), yellow rattle (*Rhinanthus*), tansy (*Tanacetum vulgare*), meadow rue (*Thalictrum*), globeflower (*Trollius europaeus*), and white veratrum (*Veratrum album*). Known ingestion of the following 28 plant taxa led to poisoning in at least one of the four animal species: sycamore (*Acer pseudoplatanus*), common box elder (*Acer negundo*), monkshood (*Aconitum napellus*), caraway (*Carum carvi*), fat hen (*Chenopodium album*), cowbane (*Cicuta virosa*), autumn crocus (*Colchicum autumnale*), hound's tongue (*Cynoglossum officinale*), bugloss (*Echium vulgare*), horsetail, marsh horsetail (*Equisetum palustre*), sweet grass (*Glyceria*), hogweed (*Heracleum sphondylium*), curled-leaved St. John's wort (*Hypericum triquetrifolium*), St. John's wort (*Hypericum perforatum*), cat's ears (*Hypochaeris radicata*), bog asphodel (*Narthecium ossifragum*), hemlock water-dropwort (*Oenanthe crocata*), parsnip (*Pastinaca sativa*), canary grass (*Phalaris*), bracken (*Pteridium aquilinium*), buttercup (*Ranunculus*), dock (*Rumex*), alpine ragwort (*Senecio alpinus*), tansy ragwort (*Senecio jacobaea*), common groundsel (*Senecio vulgaris*), black nightshade (*Solanum nigrum*), yellow oat grass (*Trisetum flavescens*), and nettle (*Urtica dioica*).

Taking into account the number of animal individuals concerned, ingestion of nine plant taxa correlated with poisoning on pastures in more than 100 individuals: sycamore (*Acer pseudoplatanus*): horses; *Cicuta virosa* (cowbane): cattle; autumn crocus (*Colchicum autumnale*): cattle and sheep; St. John's wort (*Hypericum perforatum*): sheep; bog asphodel (*Narthecium ossifragum*); bracken (*Pteridium aquilinium*): cattle and sheep; dock (*Rumex*): sheep; tansy ragwort (*Senecio jacobaea*): horses; and yellow oat grass (*Trisetum flavescens*): cattle.

3.2. Circumstances of Plant Poisoning on Pastures

In total, seven scenarios, Nos. 1–7 (categories, circumstances), were distinguished based on evidence-based (proven) cases:

1. No ingestion of poisonous plants—no poisoning (variant "avoided" of zero poisoning);
2. Ingestion of poisonous plants—yet no poisoning (variant "accepted" of zero poisoning);
3. Poisoning associated with seasonally scarce feed;

4. Poisoning associated with limited choice of feed;
5. Poisoning associated with overgrazing;
6. Poisoning associated with co-ingestion of grass;
7. Poisoning or zero poisoning under unknown circumstances.

Some plant species that caused poisoning under the circumstances numbered 3–7 (see Sections 3.2.1–3.2.7) were involved in proven zero poisoning, too, and were presented as a supplementary category (see Section 3.2.8).

When the variable "circumstance" was added to the plant-animal combination, it turned out that there were several instances of the same plant-animal combination happening under more than one circumstance (for example, tansy ragwort (*Senecio jacobaea*) and horses, associated either with overgrazing (scenario No. 5) or co-ingestion (scenario No. 6)). Therefore, the number of cases increased from 73 to 85. Zero poisoning accounted for 40% (n = 34 cases) and poisoning accounted for 60% (n = 51 cases) (Figure 1).

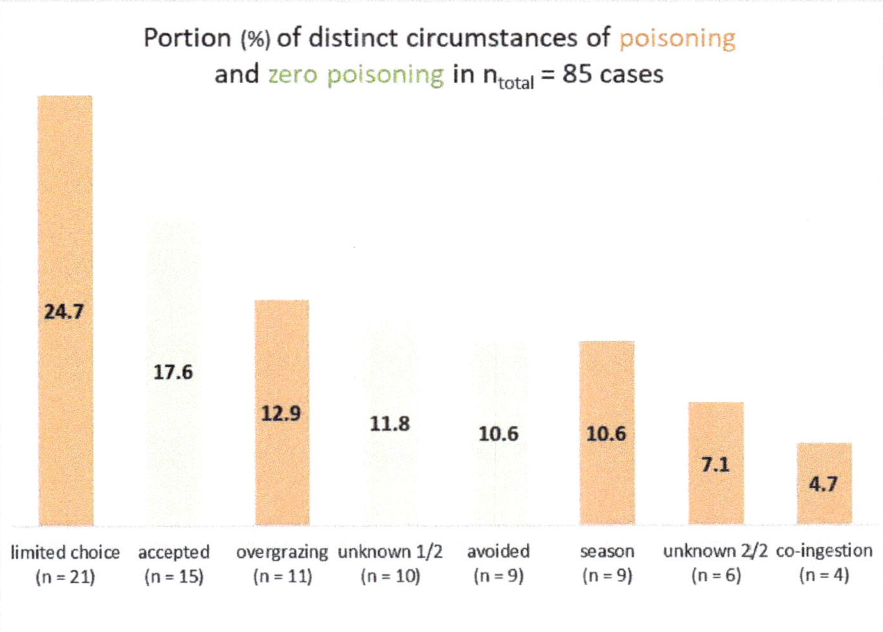

Figure 1. Distribution of poisoning (red columns) and zero poisoning (green columns) in percentage and number of cases. Only proven cases were considered (B-E; see Table 2). Abbreviations of the circumstances: "limited choice" = poisoning associated with limited choice of feed; "accepted" = no poisoning, yet ingestion of poisonous plants; "overgrazing" = feed sources were exhausted; "unknown 1/2" = circumstances of zero poisoning were not described; "avoided" = no poisoning, no ingestion of poisonous plants; "season" = poisoning associated with seasonally scarce feed; "unknown 2/2" = circumstances of poisoning were not described; "co-ingestion" = grass covered the poisonous plant that animals incidentally co-ingested.

Poisoning was most often associated with limited choice of feed plant species (24.7%) and happened the least with co-ingestion of grass (4.7%). Surprisingly, zero poisoning along with acceptance of a poisonous plant species turned out to be the second most common circumstance (17.6%). In contrast, animals avoided poisonous plants less frequently (10.6%). Both "overgrazing" (12.9%) and "seasonally scarce feed" (10.6%) only belonged to a medium-frequency class of circumstances. Representative examples of the 85 cases are presented in the following (Sections 3.2.1–3.2.7).

3.2.1. No Ingestion of Poisonous Plants—No Poisoning (Variant "Avoided" of Zero Poisoning)

Avoidance (e.g., ostensibly successful plant-animal communication) was observed for seven plant taxa in a total of nine cases (Table 3).

- Cattle
    1. "*Caltha* [*palustris*] is not eaten at all by cows" [39].
    2. *Papaver*: "[a]nimals are safe since the unpleasant odour and taste of the plants render them obnoxious (Long 1910)" [42].
- Sheep
    1. *Equisetum palustre*: "sheep were strongly selective, that means, shoots from marsh horsetail were intentionally left" [41] (translated).
- Goats
    1. *Veratrum album* was observed to be avoided by goats [26].
    2. Although goats "nibbled" creeping buttercup (*Ranunculus repens*), lesser celandine (*R. ficaria*), and bachelor's buttons (*R. aconitifolius*) [46], they avoided lesser spearwort (*R. flammula*) [26].
- Horses
    1. *Colchicum autumnale*: "[m]ost interviewees (76.2%) with horses reported that their horses neither feed on *C. autumnale* on the pasture [...]" [40].
    2. *Rhinanthus*: "[y]et there seems to be a lack of clinical data regarding actual poisonings. That might be because it's not consumed at all (have heard 3rd hand that a horse will spit it out), not enough is consumed, or perhaps mechanically little is picked up at haying" [45].

Table 3. Zero poisoning by avoidance of poisonous plants, indicated by reference numbers.

No.	Plant Species, Common Name	Plant Species, Scientific Name	Cattle	Sheep	Goats	Horses
1	Marsh marigold	*Caltha palustris*	[39]			
2	Autumn crocus	*Colchicum autumnale*				[40]
3	Marsh horsetail	*Equisetum palustre*		[41]		
4	Poppy	*Papaver*	[42]			[42]
5	Buttercup	*Ranunculus* (1)			[26]	
6	Yellow rattle	*Rhinanthus*		[43,44]		[45]
7	White veratrum	*Veratrum album*			[26]	
		Cases (n)		9		

(1) related to lesser spearwort (*Ranunculus flammula*).

3.2.2. Ingestion of Poisonous Plant—Yet no Poisoning (Variant "Accepted" of Zero Poisoning)

No evidence of health problems or poisoning was reported in 15 cases in which the animals ingested 12 poisonous plant species (Table 4). Ingestion of these species was confirmed by direct observation of grazing behavior according to the sources cited below.

- Cattle
    1. *Acer pseudoplatanus*: "Since the seedlings were not selected but always eaten along grasses and forbs, we observed their ingestion as a byproduct of grazing. [...] there is not only a direct proof of consumption on the field level by various distinct field methods but also on the chemical level as metabolites of HGA [hypoglycin A] and MCPrG (methylenecyclopropylglycine) are further detected in urine and milk samples [...]" [47].
    2. *Persicaria/Polygonum*: "[c]attle, sheep, and goats eat *Polygonum bistorta* willingly, while horses avoid it (Wyżycki 1845)" [26].

- Sheep and goats
    1. *Ranunculus*: "[e]specially sheep and goats like to nibble at poisonous plants such as Ranunculus aconitifolius, Ranunculus ficaria, Ranunculus repens. Rhinanthus spp., Rumex acetosa, Rumex alpinus [...]" [46].
    2. *Heracleum mantegazzianum*: "[b]y contrast grazing is an environmentally safe control method. Sheep grazing sustains a dense, short vegetation of forbs and grasses (Andersen, 1994). [...] For a complete eradication of Giant Hogweed a high grazing pressure is needed. Grazing by 10 sheep per ha was found to change the vegetation towards a less species rich community dominated by grazing tolerant species [...]" [49].
    3. *Tanacetum vulgare*: "[w]hat! Sheep control tansy (Tanacetum)?" [52].
    4. *Senecio jacobaea*: "[s]heep continuously preferred ragwort. The daily intake was above the currently assumed lethal dose, varying between 0.2–4.9 kg per sheep. Clinical, hematologic, and blood biochemistry parameters mostly remained within the reference limits. Initially elevated liver copper content declined over time" [51].
- Horses
    1. *Equisetum palustre*: "feeding traces showed no selectivity towards marsh horsetail" [41] (translated). This study was published in 2012. In contrast, another source from 1952 reported that "poisoning with the green plant is hardly to be observed since horses avoid the weed in pastures" [53] (translated).

Table 4. Zero poisoning when the plant species was eaten, indicated by reference numbers.

No.	Plant Species, Common Name	Plant Species, Scientific Name	Cattle	Sheep	Goats	Horses
1	Sycamore	*Acer pseudoplatanus* (juv.)	[47]			
2	Marsh horsetail	*Equisetum palustre*				[41]
3	Giant hogweed	*Heracleum mantegazzianum*		[48,49]		
4	Lousewort	*Pedicularis palustre*			[26]	
5	Smartweed/knotgrass	*Persicaria/Polygonum*	[26]	[26]	[26]	
6	Buttercup	*Ranunculus* [(1)]		[46]		
7	Yellow rattle	*Rhinanthus*			[46]	
8	Dock	*Rumex*			[46]	
9	Tansy ragwort	*Senecio jacobaea*		[50,51]		
10	Tansy	*Tanacetum vulgare*		[52]		
11	Meadow rue	*Thalictrum*	[26]			[26]
12	Globeflower	*Trollius europaeus*			[26]	
		Cases (n)		15		

[(1)] related to creeping buttercup (*Ranunculus repens*), bachelor's buttons (*R. aconitifolius*), and lesser celandine (*R. ficaria*).

3.2.3. Poisoning Associated with Seasonally Scarce Feed

Seasonally scarce feed meant either too little feed during times of drought or inadequate feed outside the core pasture season. In any case, pasture biomass was extremely reduced, causing nutrient deficiency in farm animals over long periods. Moreover, the concentration of plant secondary metabolites was higher in certain developmental stages in two plant species involved in that scenario: (1) In springtime, the concentration of hypoglycin A in *Acer pseudoplatanus* was higher in seeds and seedlings than in the adult leaves [54]. (2) In summertime, the concentration of alkaloids in *Senecio jacobaea* was higher in flowers than in the leaves [55]. Under such circumstances, nine cases of poisoning happened on pastures with nine plant species involved (Table 5).

- Cattle
    1. *Chenopodium album*: "[a]t present [1975] the west country is experiencing drought conditions. Unbelievably for us [in Great Britain], grass is in very short supply [. . .]" [59].
    2. *Pteridium aquilinium*: "[. . .] bracken [poisoning [. . .] in the year 1893 [. . .], when a drought, 'the severest of the century' (Penberthy, 1893) [in Great Britain], lasted through the spring and summer and dried up everything except the bracken to which the cattle turned as the only food available [. . .]. The disease is commonest in dry years when other herbage is scarce [. . .]" [63].
    3. *Narthecium ossifragum:* "[a] combination of factors, including the scarcity of grass caused by previous dry weather, may have contributed to this poisoning incident in County Fermanagh in the summer of 1989. [. . .] In the present outbreak of poisoning, the drier areas of the farm were grazed preferentially by the cattle [. . .]. The dry summer and aeration of the peat bog due to mechanical turf cutting operations may have contributed to the proliferation of bog asphodel on the site" [61].
    4. *Oenanthe crocca*: "[d]rought and poor grazing may also compel the cattle to seek more nutritious grass in marshy places in many of which *Oenanthe crocata* grows in abundance [. . .] The roots [. . .] are attractive to cattle because of their parsnip-like appearance and also because of their taste and odor [. . .] during a period of very dry windy weather which greatly reduced the nutritional value of the grass. Five Ayrshire heifers were found dead in a wooded, boggy area to which they had gained access by breaking through the boundary fence. The only significant finding at post-mortem examination was the presence of pieces of undigested *Oenanthe crocata* root in the rumen of each animal [. . .] [a] considerable number of roots were lying free on the shingle and all that was left of some was the growing stem about one inch high. There was no grass growing between the plants and the sea and it was concluded that the turbulence in the water during the storm had washed away the sand and gravel from the roots which were left exposed by the receding tide" [62].
    5. *Equisetum palustre*: "[...] they had found that the general condition of the cattle may deteriorate during the cool and rainy grazing periods, when they become more susceptible to *E. palustre* poisonings. In these cases there are usually other factors contributing to the sickness" [60].
- Sheep
    1. *Aconitum napellus*: A flock of 20 healthy animals was released for the first time of the year in May, when the grass had not yet grown exceptionally. Instead, the sheep fed on *Aconitum napellus* that grew within the neighboring garden, accessible to the sheep. The plants were just a few centimeters high. All seven sheep with symptoms fully recovered [58]. "The plant is not usually eaten (acrid test), and field poisoning is uncommon" [6].
    2. "*P[halaris] tuberosa* can cause death in cattle, but this occurs much more frequently in sheep on fresh growth of the plant after rain, especially at the end of a dry season in cool weather when this plant shoots more rapidly than other species [. . .]. Hungry sheep are most likely to be poisoned [. . .]" [18].
- Horses
    1. *Acer pseudoplatanus:* "[b]ecause sprouts are more common in Spring and seeds are more common in Autumn, it would appear that horses eat more seeds than sprouts either because more seeds are available or because they prefer seeds to sprouts. [. . .]. In general, Spring pasture contains more and better grass than Autumn pasture, and thus horses may have less reason to eat other feedstuff" [57].

2. *Senecio jacobaea*: "The occurrence of the case under discussion was preceded by a period of drought, and there had been but little grass on the pastures for some weeks" [65]. And another source reported: "There was one conspicuous feature common to the majority of affected farms, and particularly striking in several instances, and that was, that the grazing was of a very poor quality, the very short, good grasses and clovers being very scarce, and that plants commonly regarded as weeds were common. In general, in years with low rainfall, cases of ragwort poisoning in grazing horses increased [. . .] there were also two further peaks of incidence: first in March, when the working horses feed on hay instead of straw because of the start of the working season on the fields. Second in August when the work is over and the horses spend their free time on poor pastures" [64].

Table 5. Cases of poisoning associated with seasonally scarce feed, indicated by reference numbers.

No.	Plant Species, Common Name	Plant Species, Scientific Name	Cattle	Sheep	Goats	Horses
1	Sycamore	*Acer pseudoplatanus* (juv. and seeds)				[56,57]
2	Monkshood	*Aconitum napellus*		[58]		
3	Fat hen	*Chenopodium album*	[59]			
4	Marsh horsetail	*Equisetum palustre*	[60]			
5	Bog asphodel	*Narthecium ossifragum*	[61]			
6	Hemlock water-dropwort	*Oenanthe crocata*	[62]			
7	Canary grass	*Phalaris*		[18]		
8	Bracken	*Pteridium aquilinium*	[63]			
9	Tansy ragwort	*Senecio jacobaea*				[64,65]
		Cases (n)		9		

Number of affected animal individuals (n): beige: n < 10; orange: n > 100.

3.2.4. Poisoning Associated with Limited Choice of Feed

In contrast to circumstances of seasonally sparse grass, limited choice meant that the biomass of 1 of the following 17 plant species dominated in 21 cases (Table 6).

- Cattle
    1. *Trisetum flavescens*: This species made up the largest portion of biomass, between 18 and 25%, in pastures with a high diversity (10 grass and 21 herbaceous species) [82]. Except for *Anthriscus* (5–18%), *Dactylis* (8–11%), and *Trifolium* (3–18%), all other species reached biomasses between 1 and 8% [82]. In contrast, this grass species was missing on extensively exploited pastures without any fertilization and animal movements as well as on intensively utilized pastures with abundant fertilization and frequent movement of cattle [81]. "On pastures on the alp, the symptoms gradually diminish, however, appear more intensive in the following winter. There is little yellow oat grass on the alp pastures, yet much in the dales" [83] (translated).
    2. *Cicuta virosa*: "[. . .] [on] low-lying, marshy land where *C. virosa* grows abundantly. The authors stated that cattle find the plant palatable and are attracted by its smell. "[. . .]. Cattle should be prevented from grazing in the neighbourhood of water reservoirs on low-lying pastures especially during drought" [67].
    3. *Colchicum autumnale*: "the vegetation consisted mainly of rushes [Juncus] and coverage of *Colchicum* reached 40% on 50–100 m^2" [68] (translated).
    4. *Echium vulgare*: "[t]en 1-year-old fighting bulls died between October and March from a herd of 700 animals [. . .]. All animals had grazed in the pastureland [. . .]. Close inspection of the pastureland where animals were grazing revealed large quantities of *Echium vulgare* (80%) and *Senecio vulgaris* (15%) [. . .]" [70].
    5. *Narthecium ossifragum*: "[d]uring the summer of 1992, an apparently new disease with renal failure as its dominating sign occurred in grazing cattle in western Norway (Fl¢yen et al., 1993) The animals were grazing uncultivated pastures and *Narthecium ossifragum* was a common plant in all the areas where the disease

occurred. Two hundred and thirty-two animals suffered from the disease and 137 died" [84].

6. *Senecio jacobaea*: "[s]everal inspections of these pastures in June and July 1997 revealed that *S[enecio] alpinus* was very common" [78]. "Ten out of 75 calves died or had to be slaughtered after they had grazed on wasteland carrying a heavy growth of *S. jacobaea*" [80].
7. *Cynoglossum officinale*: "[t]owards the end of April, 1959, a herd of 23 Friesian cows in various stages of pregnancy gained access to an area of waste land and had been grazing there for several hours before their escape was discovered. [...] It was noticed immediately that the cattle were grazing an area heavily contaminated with hound's tongue or dog's tongue in the leafy stage of growth [...]. Heavy cropping of the plants had taken place. [...] The rumen contained large quantities of the leaves of the hound's tongue [...]" [69].
8. *Hypericum crispum* "is very eagerly eaten by herbivores in Turkey; therefore, one cannot stop husbandry animals from ingesting this plant in large quantities during each summer on fields, ditches, and dried meadows as the only grazing sites" [72] (translated).
9. *Ranunculus*: "A cow on a pasture with a heavy crop of buttercups (*Ranunculus bulbosus*) became ill, with salivation, coughing, restlessness and rigors. It recovered the next day" [75]. "After having been housed and given good feed, cows in late pregnancy were turned out onto fields where the pasture was poor and contained buttercups (*Ranunculus sp.*). Within 14 days clinical signs including tympany, diarrhoea with blood and mucus, rapid pulse and difficult breathing developed and 5 animals died or were killed shortly before or after calving, despite symptomatic treatment. Consistent postmortem findings were intestinal haemorrhages, fatty liver and fat infiltration of kidneys" [76]. "[...] two heifers, apparently developed a taste for *R. sceleratus*. Having just recovered from acute poisoning, they returned to the same place and began to eat the buttercups again; they had to be removed from the field for their own safety" [18].

- Sheep
    1. *Pteridium aquilinum*: "[s]heep are sometimes so reluctant that they seem ready to starve rather than eat the bracken. [...] [T]he sheep had little choice but heather or bracken to eat. Thirteen sheep had been lost in a few weeks in a flock of only 43 ewes" [63]. "Since 1950 cases of bracken poisoning with sheep increased, while the frequency of burning down heather decreased, leading to a fiber-rich herbage that the sheep avoided in favor for bracken as only available feed." [63] There were "[...] only heather and a few other plants such as bilberry while the slopes are heavily covered with bracken concealing a poor carpet of grass. The observation that is experimentally it is very difficult to persuade sheep to eat bracken fits in the general impression that sheep avoid it" [63].
    2. *Chenopodium album*: three out "[o]f 40 ewes transferred after shearing to a sown pasture in which the grass had failed and *C. album* predominated, 1 died and 2 had to be slaughtered" [66].
    3. *Narthecium ossifragum*: "[p]hotosensitization occurs annually in the Blackface sheep on three hill farms in Perthshire. [...] Shepherds are insistent that only lambs become affected. [...] An examination of the pastures was carried out by Dr. D. Martin of the West of Scotland College of Agriculture. He reported that the pasture was a typical hill sward. *Molinia* was dominant and *Narthecium ossifragum* was present. When the wetter parts of the hill were examined closely, *N. ossifragum*, the bog asphodel, was seen to be more common than originally thought. The spike-like leaves protruded through tufts of *Sphagnum* and many of the tips had been bitten off" [74].
    4. *Hypericum perforatum*: "in a flock with 700 ewes, kept within a large paddock, almost 200 animals were fallen sick within three days. On this extensive grazing

system, a large portion of *Hypericum perforatum* was present. While being tented, St. John's wort is being avoided or eaten in small amounts only. In contrast, keeping sheep on paddocks showed that the animals had ingested St. John's wort down to the stems" [73] (translated).

5. *Rumex acetosella*: "in a flock with more than 600 ewes, a chronic disease was observed, marked by oedematous heads and throats as well as by increasing emaciation. It has been observed that *Rumex acetosella* was present to a great extent on the pastures and made up more than 50% of the vegetation on some spots" [73] (translated).

- Horses

1. *Heracleum sphondylium*: "[t]he animal [horse] had been moved 10 days previously to a pasture that had not been grazed by animals for a number of years. On inspection, around 50 per cent of the pasture consisted of hogweed plants [. . .]" [71].
2. *Senecio alpinus*: "during inspection of the associated mountain pasture, apart from a species-poor vegetation of grasses, a dominating contamination with *S. alpinus* on a large scale was found within a partly tall-grown vegetation on a strongly eroded, endangered, northern exposed site. The moist soil had been fertilized through intensive grazing activities. Alpine ragwort, densely growing in groups, had been increasingly eaten. In contrast, single plants remained often untouched" [79] (translated).

Table 6. Incidents of poisoning associated with limited choice of feed plant species, indicated by reference numbers.

No.	Plant Species, Common Name	Plant Species	Cattle	Sheep	Goats	Horses
1	Fat hen	*Chenopodium album*	[66]	[66]		
2	Cowbane	*Cicuta virosa*	[67]			
3	Autumn crocus	*Colchicum autumnale*	[68]			
4	Hound's tongue	*Cynoglossum officinale*	[69]			
5	Bugloss	*Echium vulgare*	[70]			
6	Marsh horsetail	*Equisetum palustre*				
7	Giant hogweed	*Heracleum sphondylium*				[71]
8	Curled-leaved St. John's wort	*Hypericum triquetrifolium*	[72]	[72]		[72]
9	St. John's wort	*Hypericum perforatum*		[73]		
10	Bog asphodel	*Narthecium ossifragum*		[74]		
11	Bracken	*Pteridium aquilinium*		[63]		
12	Buttercup	*Ranunculus*	[75,76]			[77]
13	Dock	*Rumex*		[73]		
14	Alpine ragwort	*Senecio alpinus*	[78]			[79]
15	Tansy ragwort	*Senecio jacobaea*	[80]			
16	Common groundsel	*Senecio vulgaris*	[70]			
17	Yellow oat grass	*Trisetum flavescens*	[81]			
		Cases (n)		21		

Number of affected animal individuals (n): beige: n < 10; light red: n > 10; orange: n > 100.

3.2.5. Poisoning Associated with Overgrazing

"Overgrazing" meant that feed sources were exhausted because pastured animals had ingested the whole biomass of acceptable plants. Under such circumstances, they had started to feed on 10 poisonous plant species in 11 cases (Table 7).

- Cattle

1. *Aconitum napellus*: "One single case series shows a repeated, although exceptional, individual behavior towards monkshood (*Aconitum napellus*). Each year from 1995 to 1999, exactly between 9 July and 15 July, 1 heifer died out of a herd of 80 pastured animals. On the 20th of July, each year, the herds were brought to

another pasture. "Toxicosis usually occurs during a 'toxic window' or when little else is available to eat" [86].

2. *Glyceria aquatica*: "one cow, found dead, unfortunately had ingested a fair amount of sweet grass (*Glyceria aquatica*) that was present at the riverbank since there was a shortage of grass that has become rare on the meadow" [88] (translated). "Starting point was the intake of reed sweet grass by the [pregnant] heifers, ingested out of particular need for energy. Adapted to wet sites, this grass species possesses broad, fleshy leaves and forms lush vegetation that reaches a height of some 200 cm [...]. Also due to its low fibre (3.89%) and high protein content (6.46%) with a dry matter content as low as 23.2% it represents an attractive food before flowering" [87].
3. *Pteridium aquilinum*: "There is no doubt that bovine bracken poisoning causes serious losses to hill farmers in some years —when the ecological conditions are conducive to its appearance. I have seen a Welsh farmer lose 22 pedigree Friesian heifers—his whole herd replacements valued at £2000 in 1950, with it. [...] he had been asked by his own son when passing a badly bracken infested hill, 'why don't those cattle get bracken poisoning?' And he had replied, 'John, I have never had bracken poison there and cattle have grazed on that hill all my years here.' That night a message had been received about the same cattle saying there was suspected bracken poisoning. Sixty-six heifers were grazing on that hill and 1 had died. The remainder were immediately moved on to another field and every animal was treated with batylalcohol as a preventative: some animals which became ill were treated again with batyl alcohol as a curative but with absolutely no effect. The result was that out of 66 animals grazing on that hill 54 died and some died from bracken poisoning more than a month after being removed from the hill." [90]. In another case, dietetic reasons were suggested: "[...] an exceptional outbreak in which some 30 cattle died when several hundred were placed on a new[-seeded] pasture in sustained wet weather [...]. The cattle here left the young lush grass and deliberately sought the young bracken shoots as if in search of some change, possibly a more fibrous food" [44].
4. *Senecio jacobaea*: "I generally observe five to six cases every year in animals [cattle] commonly from about 12 to 15 months old. It is usual for one case only to occur on any farm in this district, but occasionally two cases of the disease may occur, and in very occasional outbreaks several animals may be affected. [...] In this instance the animal [cattle] was allowed to graze for a few hours each day in pasture which was very short of grass, it being in the very early spring. Most of my observations are confined to the occurrence of the disease in cattle on pasture during the spring and summer seasons. The grazing is always short and land frequently overstocked" [64].

- Goats
  1. *Urtica dioica*: "in midsummer on an almost bare ground, a flock of goats was kept on some 5000 m^2. One individual showed swollen parts of the mandible. Parasitosis was excluded since feces control revealed no infestation. After the ground had been checked botanically, the owner reported that this goat regularly consumed the nettles as the only plant species still available in large amount. After the goats had been given access to a new pasture, the symptoms abated" (translated; personal communication) [37].
- Horses
  1. *Ranunculus*: "[o]n a pasture consisting mainly of the buttercups Ranunculus acris and Ranunculus sceleratus, a 4-year-old horse developed paresis of the hind quarters" [77].
  2. *Senecio jacobaea*: "I have abundant evidence that over-stocking and under-feeding have the greatest possible influence in the production of the disease [with draught

horses]" [64]. "I have observed the disease in two animals, one a 5-months-old foal and the other an aged gelding, which had been grazed on three fields of temporary grass in September, 1928, the foal having died a few weeks before its older companion. The dam of the foal and a number of young cattle had also been grazed on these fields, which were small in acreage with grass very short and full of weeds, the most conspicuous being the first and second year old stages of the common ragwort plant [. . .]. The dam is still alive, and has so far presented no symptoms indicative of liver cirrhosis" [64].

3. *Acer negundo*: "the meadow [in mid-April] was very bare and I had recommended to put the ponies calmly onto the bare meadow since they were well fed" (personal communication) [38] (translated).
4. *Hypochaeris radicata*: "[f]ood scarcity along with a relatively increased coverage of *H. radicata* (up to 20%) could explain the appearance of the disease in the late season on the grazing areas investigated" [89].
5. *Pastinaca sativa*: "on a 2-hecatare pasture, having been grazed for years by some 10 horses without problems, one horse with a white-skinned mouth developed severe photodermatitis in August. At that time, the grasses had been eaten down to a lawn-like vegetation of 3 cm in parts, from which plentiful fruiting shoots of parsnip protruded. On a representative area of 100 m^2, 548 apical parts of shoots were missing, corresponding to 91% of the whole number of shoots counted. Since the remaining basal parts of the shoots showed significant signs of damage through grazing, many horses had probably ingested the fruiting parsnips; however, the only white-skinned animal had been suffering from photodermatitis the first time" (translated; personal communication) [36].

Table 7. Incidents of poisoning associated with overgrazing, indicated by reference numbers.

No.	Plant Species, Common Name	Plant Species, Scientific Name	Cattle	Sheep	Goats	Horses
1	Sycamore	*Acer pseudoplatanus* (juv. and seeds)				[85]
2	Common box elder	*Acer negundo* (juv.)				[38]
3	Monkshood	*Aconitum napellus*	[86]			
4	Parsnip	*Carum carvi*	[87]			
5	Sweet grass	*Glyceria*	[87,88]			
6	Cat's ears	*Hypochaeris radicata*				[89]
7	Parsnip	*Pastinaca sativa*				[36]
8	Bracken	*Pteridium aquilinium*	[90]			
9	Tansy ragwort	*Senecio jacobaea*	[64]			[64]
10	Nettle	*Urtica dioica*			[37]	
		Cases (n)		11		

Number of affected animal individuals (n): beige: n < 10; light red: n > 10; orange: n > 100.

3.2.6. Poisoning Associated with Co-Ingestion of Grass

Poisoning associated with co-ingestion of grass occurred in four cases when grasses covered four poisonous plant species that animals incidentally co-ingested (Table 8).

- Cattle

1. *Equisetum palustre*: "[i]nevitably, the heifers ingested not only reed sweet grass but marshhorse tail, too, which was growing in between with a relative biomass of 5%. Thus, they had no choice to avoid a plant that cattle normally do otherwise. The haemorhagic enteritis that the young cows developed, i.e., the symptoms of palustrin poisoning, is congruent to symptoms in a feeding trial: Cattle reacted to 34.7 g TS marsh horsetail/100 kg BM/d with diarrohea [5]" [87].
2. *Senecio alpinus*: "moreover, cows are not able to distinguish the still young plants from freshly growing grass in spring or after the first hay cut, and do ingest them (Duby, 1975)" [91] (translated).

- Horses
    1. *Acer pseudoplatanus*: "[. . .] the authors documented purely the spontaneous ingestion that happened rather unintentionally with stage 2 seedlings, in particular when the horse seemed to be attracted by lush grass wherein these seedlings were more or less hidden. The latter was also true for stage 4 seedlings. Although during the whole observation time of nearly two months (8 April to 31 May), there was an increasing presence of stage 4 seedlings [. . .] and a corresponding higher and denser grown grass that might have distracted the horses from stage 4 seedlings, the young woody plants were almost invisible in the thick lush vegetation typical in spring, yet still as short as the grass layer" [54].
    2. *Senecio jacobaea*: "[i]n the outbreak described here, the clinical and laboratory findings in exposed horses are presented together with epidemiological evidence that pasture rather than hay was the principal source of the alkaloids. [. . .] Examination of the pasture in the spring of 1982 revealed a patchy, but in places very heavy, growth of young ragwort plants at the same height as the sward. [. . .] Groundsel (see *vulgaris*) was also present in moderate amounts. [. . .] Little ragwort was observed on these pastures in previous seasons and the finding of many small ragwort plants emerging with the grass in the spring suggests a recent heavy establishment of the plant in this pasture, horses perhaps grazing it inadvertently with the grass [. . .]" [92].

Table 8. Incidents of poisoning related to co-ingestion of grass, indicated by reference numbers.

No.	Plant Species, Common Name	Plant Species, Scientific Name	Cattle	Sheep	Goats	Horses
1	Sycamore	*Acer pseudoplatanus* (juv.)				[54]
2	Marsh horsetail	*Equisetum palustre*	[87]			
3	Alpine ragwort	*Senecio alpinus*	[91]			
4	Tansy ragwort	*Senecio jacobaea*				[92]
		Cases (n)			4	

Number of affected animal individuals (n): beige: n < 10; light red: n > 10; orange: n > 100.

In all cases of co-ingestion, animals were in particular need of energy either due to the season (spring and fall) or due to pregnancy (Figure 2).

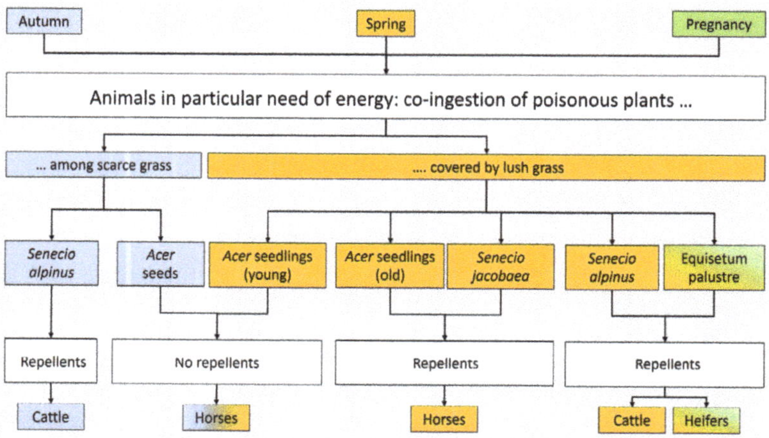

Figure 2. Co-ingestion as a circumstance in distinct plant and animal species. Repellents are assumed on the basis of observations of grazing husbandry animals in the case of older *Acer* seedlings (phenols), *Senecio* (sesquiterpenes), and *Equisetum* (silicate). For references of the cases, see Table 8.

3.2.7. Poisoning or Zero Poisoning under Unknown Circumstances

In 16 cases, information was missing as to the circumstances of whether the plant had been eaten or not (cases of zero poisoning) or as to the conditions of the pasture when the plant had been consumed (cases of poisoning). Thirteen plant species were involved in ten cases of zero poisoning (with nine plant species) and six cases of poisoning (with four plant species) (Table 9).

- Cattle
  1. *Cicuta virosa*: "[a] neighbouring area had been dressed with ammonia fertilizer which could have disturbed the sense of smell of the cattle. Possibly the cattle had mistaken the plant for a root crop" [95].
  2. *Colchicum autumnale*: "11 cases of poisoning: (1) June: in trenches many *Colchicum* plants, one dead animal; (2) May: Half of the pasture covered with *Colchicum*, one dead, four sick; (3) month not indicated: six dead animals on a pasture with grazed *Colchicum* plants; (4) June: four dead animals on a pasture with grazed *Colchicum* plants; (5) month and pasture conditions not indicated: one sick animal; (6) month not indicated: four dead animals with *Colchicum* in the rumen; (7) May: three sick bulls on a pasture where some *Colchicum* plants were grazed; (8) month and pasture conditions not indicated: four sick cows; (9): month and pasture conditions not indicated: one sick cow; (10) September: one cow, new in the region, ate *Colchicum* flowers [in the original "veilleuses de nuit"]; (11) Spring: three dead animals on a pasture with abundant *Colchicum* plants" [20] (translated).
  3. *Narthecium ossifragum*: "The disease occurred on pasture which had been limed at the end of July 1989 and immediately stocked with 25 suckler cows and their calves, aged two to three months. By mid-August, 15 of the cows were in poor body condition but continued to graze. Over the following two weeks these cows became anorexic and rapidly lost body condition; 11 became recumbent and either died or were euthanased on welfare grounds, but the other four recovered" [61].
- Sheep
  1. *Colchicum autumnale*: "[a]lmost all the animals in a large nomadic herd of sheep developed clinical signs of intestinal irritation and intermittent diarrhoea with mucus and sometimes blood, as they crossed a swampy region densely populated with meadow saffron (*C. autumnale*). The sheep were exhausted and hungry and 2 that died were depressed and had difficult, noisy breathing" [97]. "A high annual incidence of *Colchicum autumnale* (meadow saffron) poisoning usually occurs in May to July in lambs grazed on the highlands of the Central Tien-Shan in Kirghizia" [19].
  2. *Trisetum flavescens*: "[a] [Kentucky blue grass] *Poa pratensis*, [rough meadow grass] *P. trivialis*, [rye grass] *Lolium perenne*, [meadow fescue] *Festuca pratensis*, *Trisetum flavescens*, [meadow oat grass]; *Helictochloa pratensis*, *Avenula pratensis*, [cock's foot] *Dactylis glomerata* and [white clover] *Trifolium repens* sward at 680–720 m alt. was rotationally grazed by a mixture of heifers and sheep (2 livestock units/ha) or by heifers and sheep alternately. Alternate grazing produced greater av. daily liveweight gains in cattle (an additional 105 g/d) whereas mixed grazing improved sheep liveweight gains (an additional 8 g/d). DM herbage yields under both grazing systems were similar at 8 t/ha" [105].
- Goats
  1. *Solanum nigrum*: "[o]n examining the pasture, considerable quantities of a weed were found, and it was obvious that it had been freshly eaten" [103].

- Horses
    1. *Senecio jacobaea*: "on the basis of a pathological-histological finding, a poisoning with common ragwort has been assumed. Larger quantities of this plant were found on the pasture, and the hay was also obtained from the affected meadows" [100] (translated). "The animals concerned were all riding horses with an average age of 15 years. The pastures were of bad quality with a large amount of common ragwort. There was concern that an estimated four horses had ingested common ragwort on the pasture" [101] (translated).
    2. *Cicuta virosa*: "[p]ortions of cowbane (*Cicuta virosa*) stems were found in their stomachs, and the bases of chewed-off plants were found in the field" [96].

Table 9. Incidents of zero poisoning and poisoning under unknown circumstances, indicated by reference numbers.

No.	Plant Species, Common Name	Plant Species, Scientific Name	Cattle	Sheep	Goats	Horses
1	Marsh marigold	*Caltha palustris*				[93]
2	Fat hen	*Chenopodium album*				[94]
3	Cowbane	*Cicuta virosa*	[95]			[96]
4	Autumn crocus	*Colchicum autumnale*	[20]	[19,97]		
5	Horseshoe vetch	*Hippocrepis comosa*	[98]	[98]		
6	Smartweed/knotgrass	*Persicaria/Polygonum*				[26]
7	Bracken	*Pteridium aquilinium*			[28]	
8	Yellow rattle	*Rhinanthus*	[99]			
9	Dock	*Rumex*				[94]
10	Tansy ragwort	*Senecio jacobaea*				[100–102]
11	Black nightshade	*Solanum nigrum*			[103]	
12	Yellow oat grass	*Trisetum flavescens*		[104,105]		
13	Nettle	*Urtica dioica*	[106]			
		Cases (n)			16	

Number of affected animal individuals (n): dark green: n = 0; beige: n < 10; orange: n > 100.

3.2.8. Cases of Proven Zero Poisoning with Plant Species that Have Caused Poisoning in Other Cases

Cases of proven zero poisoning of the whole animal population with plants that had yet to cause poisoning in other cases were thus not counted as zero poisoning but as poisoning. Therefore, the following five cases did not appear in the above-presented figures and tables.

- *Equisetum palustre* and cattle: "Thus, the animals first seek selectively seek out the areas less covered with marsh horsetail to feed. Marsh horsetails were often found on or in the avoided tufted sedge. However, it was not possible to determine whether the animals had selectively eaten the marsh horsetail in particular" [41] (translated).
    1. *Senecio jacobaea* and cattle: "A herd of some 15 cattle grazed on pastures where ragwort was dominating and a control group of another 15 animals on a pasture without ragwort. After one year, no individuals showed any elevated liver enzymes or clinical symptoms" (translated) [107].
    2. *Senecio alpinus* and cattle: "unquestionably, the animals grazing on the pasture do not take this plant voluntarily, as long as there is enough feed. It could be clearly observed during the inspection of the farm in the canton Schwyz that tufts of alpine ragwort stood up tall above the other plants in the meadow and were neither partly nor completely ingested" [91] (translated).
    3. *Hypericum* and sheep: as already cited (Section 3.2.4): "while being tended, St. John's wort is being avoided or eaten in small amounts only" [73] (translated).
    4. *Senecio jacobaea* and horses: "An example of yearlong grazing of horses on pastures with ragwort revealed that the animals, both young and adult ones, did have co-ingested the plant in small amounts, but did not fall ill" (translated) [108].

4. Discussion

The aim of this review is (1) to document and evaluate the circumstances of pasture poisoning in the most common husbandry species, cattle, sheep, goats, and horses, and (2) to present a corresponding checklist of poisonous pasture plant species in Central Europe.

4.1. Hunger as Driving Force in Cases of Pasture Poisoning

Apart from only six cases of poisoning under unknown circumstances (7.1%; see Figure 1) as well as one case of assumed dietetic reasons (cattle and *Pteridium*; see point 3 in Section 3.2.5), the data revealed that severe health problems or fatal poisoning due to poisonous plant species evolved practically always due to hunger as a natural, indeed banal [109], consequence of ingesting everything available regardless of the quality. This finding was argued as early as 1926: "The conditions under which animals are tempted to eat poisonous plants are of considerable importance. We have still much to learn on this subject, and the greater part of the necessary information can be collected and recorded only by workers in the field. Excessive hunger is doubtless the main cause of the consumption of poisonous plants, and that arises from a relative scarcity of other foodstuffs, which is due most commonly to particular weather conditions, drought especially. Placing too many animals on a given grazing area will obviously have the same effect of inducing a shortage of other and more palatable foodstuffs. Animals which have been travelling are particularly liable to eat poisonous plants; probably this is often merely a matter of hunger. [. . .] the first step should be to determine whether the plant will cause harm when ingested by animals, and whether any animal is likely ever to consume an effective quantity under any combination of natural conditions" [110]. Most often the starting point is insufficient feed choice, while the number of cases of overgrazing seems surprisingly low against such a background. However, the single kinds of circumstances are hardly to be distinguished: seasonally scarce feed would result in overgrazing which leads in turn to insufficient feed choice or co-ingestion of poisonous plants. Examples such as the following one reveal that pasture poisoning is a multivariate phenomenon, too: "[...] they had found that the general condition of the cattle may deteriorate during the cool and rainy grazing periods, when they become more susceptible to [marsh horsetail] *E[quisetum] palustre* poisonings. In these cases there are usually other factors contributing to the sickness" [60].

Co-ingestion is an exceptional phenomenon in various regards. It is not only an extreme symptom of hunger, but also always associated with a special physiological need for energy of the animals involved. In contrast to other scenarios, not scarce but lush grass turns out to be the problem in all variants of this kind of poisoning, except in *Acer* seeds that most often fall on sparse grass in the fall. While ingesting the lush grass, animals obviously accept or do not fully sense repellents in ragwort (*Senecio*), marsh horsetail (*Equisetum palustre*), or juvenile seedlings of sycamore (*Acer pseudoplatanus*) hidden in the nutritious feed, although at least horses can distinguish between gradients of repellents. Horses ingested 19.1% of juvenile seedlings with 1.65 mg total phenolics/g fresh weight (FW), yet only 5.46% of older seedlings with 8.48 mg total phenolics/g FW [54]. Hunger would lower any tolerance towards unpalatable plants in order to survive, in particular in distressed animals: "[Blister buttercup] *R[anunculus] sceleratus* is most likely to give rise to problems, owing to its occurrence in damp places where it can provide lush vegetation when the rest of the pasture is comparatively bare [or covered with unpalatable sedges]" [18]. Even for monkshood (*Aconitum*), the species with one of the most potent toxins in plants in Europe [111], the tolerance level may be lowered; however, the tolerance level differs individually, as the corresponding case reports in sheep (see Section 3.2.5) and cattle prove (see Section 3.2.3). Only a few animals of the whole flock were affected (7 of 20 sheep, all recovered; 1 of 80 heifers died). These data confirm that the threshold for the tolerance of poisonous plants is fundamentally individual, just as the tolerance of hunger or the threshold of perception of repellents and other plant signals is individually pronounced [112]. Hunger, as described above, obligatorily interferes with any plant-animal communication and changes animal behavior, resulting in a dysfunctional acceptance of unpalatable plants in certain animal individuals or in entire herds. Zero poisoning with plants such as horse marshtail (*Equisetum palustre*), tansy ragwort (*Senecio jacobaea*), and alpine ragwort (*Senecio alpinus*) in cattle or St. John's wort (*Hypericum perforatum*) in sheep, causing poisoning under other circumstances, shows that poisonous plants are not automatically but potentially poisonous, depending not upon the toxins but on the circumstances.

According to the data of publications, poisoning of entire cattle herds with cowbane (*Cicuta*) in 1955, autumn crocus (*Colchicum*) in 1968 and 1975, bog asphodel (*Narthecium*) in 1988, bracken (*Pteridium*) in 1964 and 1965, and yellow oat grass (*Trisetum*) in 1980 is a historical finding in Central Europe. Pastured cattle are becoming scarcer [113], while the number of horses on paddocks has risen steadily during the last 50 years [114]. As an interesting fact, poisoning of large horse populations has occurred two times: in cart horses in the 1930s with tansy ragwort (*Senecio jacobaea*) and in the present millennium in pleasure horses with sycamore (*Acer pseudoplatanus*).

4.2. Do Poisonous Plants in Pastures Communicate Their Toxicity?

The proportion of 40% zero poisoning in all evidence-based cases is surprisingly high in a database where cases of zero poisoning are underrepresented due to their lack of clinical relevance. This figure shows that plants and animals on pastures can co-exist to a large extent. Obviously, plants with plant secondary metabolites have successfully mediated their message of being harmful or not to grazing animals even within the confined conditions of fenced pastures.

Zero poisoning happens both via avoidance and acceptance and is based on pre- and post-ingestive mechanisms [1], mentioned in the introduction. However, as late as 1952, one could explain zero poisoning in horses of pastures with *Equisetum* only with avoidance. Physiological (safe) ingestion of poisonous plants is enabled by both behavioral and physiological mechanisms. First, the kind and amount of ingested poisonous plants depends upon species-appropriate socialization [115] and sustainable experiences [116] of young animals. Second, the species-specific tolerance relies on the detoxification mechanism. For example, cattle feed on *Acer* seedlings due to their tolerance of hypoglycin A [117], and sheep tolerate pyrrolizidine alkaloids of their preferred feed plant *Senecio jacobaea* [118]. In goats grazing *Rumex*, a mechanism of adaptation to oxalate digestion seems to be responsible [119]. However, in the case of horses grazing marsh horsetail (*Equisetum palustre*), we know neither the tolerance nor the detoxification level, nor the pre-ingestive pattern of sensation, and even not whether the responsible toxin or repellent is palustrine, thiaminase, silicate, or another compound [120] and whether it acts as a kind of "feeding brake". The same is true for most of the plant species, including those with potent toxins, such as giant hogweed (*Heracleum mantegazzianum*), tansy (*Tanacetum vulgare*) (both eaten by sheep), and white veratrum (*Veratrum album*) (avoided by goats). The variety of plant secondary metabolites is astonishing and their functions in particular for vertebrate herbivores are widely unknown: Did the 120 phenolic compounds in older sycamore seedlings on horse paddocks [54] contribute more effectively to the higher rate of avoidance by horses than the 82 compounds in young seedlings? Is there any synergistic effect? Are there certain compounds more relevant than others? At what concentration is a single compound perceived?

In contrast, assumed (not evidence-based) cases of zero poisoning probably cannot be exclusively traced back to successful plant-animal communication for two reasons: First, some combinations are unlikely to happen: goats would avoid all plants on wet sites such as marsh marigold (*Caltha*), cowbane (*Cicuta*), and sweet grass (*Glyceria*) because they simply avoid the water-flooded sites of those plants. Second, zero poisoning probably includes overlooked cases of asymptomatic or subclinical poisoning, too: "[...] it is difficult to assess the full effects of plant poisonings, as the toxic compounds may cause only indistinct signs such as mild digestive disturbances or reduced fertility, that pass more or less unrecognized" [1]. Moreover, the tolerance level towards hunger or repellents might be highly variable and individual, as summarized as early as 1947 "[o]ur knowledge of the whims and reactions of the grazing animal—his psychology in short—is limited and elementary in the extreme [...]" [121].

On the basis of the circumstances analyzed here, as long as there is feed choice, even scarce feed is not a problem: "[i]t is remarkable that I have never observed a case of the [liver] disease [through tansy ragwort (*Senecio jacobaea*) in horses] on a mountain farm, where tillage is naturally very limited, where the bulk of the grazing consists of permanent virgin pasture, and where the hay commonly used is of the meadow variety" [64]. An unspectacular co-existence is typical for numerous popular plant species with plant secondary metabolites such as sour dock (*Rumex acetosa*). The plant contains oxalic acid that may bind calcium in the blood, a process that can lead to hypocalcemia [122]. Although sour dock (*Rumex acetosa*) grows on almost every pasture in Central Europe, no cases of poisoning are known. Another prominent example is ground ivy (*Glechoma hederacea*), which is poisonous to horses in the case of forced feeding [123] but obviously never eaten on pastures since no report exists (assumed zero poisoning) [109].

4.3. Checklist of Poisonous Plants as a Tool for Risk Management

The checklist comprises 52 species and might not be complete when one considers the species richness of European grasslands. Its advantage is both the incorporation of suspicious plant species that were revealed to be no problem for grazing animals and the exclusively empirical basis due to documented or mainly peer-reviewed case reports. In other words: The checklist contains cases of symptomatic poisoning (negative list) as well as cases of zero poisoning (positive list). Moreover, the checklist includes cases of assumed zero poisoning under the premise that no reference—neither of poisoning nor of zero poisoning—is available. The inclusion of assumed and proven zero poisoning along with proven poisoning is the only possibility for completely documenting the state of the art on an empirical basis of evidence-based cases. This may be of practical relevance, too, since funding programs of the European Union would not only encourage the grazing of husbandry animals but also promote floristic diversity on grassland. In this context, farmers encounter the dilemma that with diversity, the number of poisonous plants increases as well [124]. However, with the availability of data on the circumstances of cases, farmers can avoid accidental pasture poisoning in the case of certain plants (negative checklist) and may evaluate the risks of certain other plants (positive checklist). There might also be potential benefits of poisonous plants (weeds) on pastures since "[m]any poisonous plants in small doses are medicinal, such as [red knees] *Persicaria hydropiper*, [columbine meadow rue] *Thalictrum aquilegiifolium* and others" [125]. One can even speculate that these unpalatable plants such as poisonous species and weeds might stimulate movement in pleasure horses since the the animals will need to walk further in search of better feed—as long as there is something better.

5. Conclusions

"The viper, though it kills with it, does not deserve to be blamed for the poison it carries, as it is a gift of nature" [126]. From the point of view of the plant, poisonous plant species may become dysfunctionally toxic under two circumstances: if non-voluntarily ingested on pastures with scarce feed plants or if unintentionally co-ingested by animals in special need of energy.

These findings are all the more relevant as floristic diversity became a goal of modern grassland management during the past decade [113]. In particular, the possibility of free feed choice for husbandry animals has been recently emphasized, too [127]. Floristic diversity along with both provision of adequate feed and avoiding overgrazing would not only offer feed choice between nutritive, dietetic, and poisonous plants but also facilitate successful plant-animal co-existence on pastures. This meta-study of the circumstances of pasture poisoning reveals that plant-animal communication inevitably fails most often simply due to feed deficiency. An individual animal might utterly perceive the communication of toxicity by the plant species but be forced to ignore the message due to a limited choice of feed options. This is different from a failure in communication.

Although there is still an insufficient understanding of the role of secondary plant metabolites for herbivorous vertebrates, the question of whether poisonous plants on pastures communicate their toxicity in general can empirically be answered in the affirmative.

Supplementary Materials: The following supporting information can be downloaded at https://www.mdpi.com/article/10.3390/ani13243795/s1, Table S1. Checklist of poisonous plants in pastures in Central Europe and references. References [15,18–20,26,28,34,36–52,54,56,58–81,85–106,109,120,122,128–140] are cited in the supplementary materials.

Funding: The APC was funded by Deutsche Forschungsgemeinschaft (DFG), grant number 491094227.

Institutional Review Board Statement: Not applicable.

Informed Consent Statement: Not applicable.

Data Availability Statement: The raw data are deposited at the repository of the Institute of Animal Nutrition of the University of Veterinary Medicine Hannover and will be readily given on demand.

Acknowledgments: I would like to thank the librarians from the University of Veterinary Medicine Hannover, Foundation. The colleagues managed to detect hidden printed issues, often insufficiently mentioned in the cited references. Moreover, I am grateful to the three anonymous reviewers, and one of them in particular, for their outstanding, thorough, and stimulating work. Their critical focus on the manuscript has improved basic aspects, such as the state of the art, definitions, and conclusions. Finally, my sincere thanks go to Frances Sherwood-Brock from the Editorial Office of the University of Veterinary Medicine Hannover.

Conflicts of Interest: The author declares no conflict of interest.

References

1. Ingebrigtsen, K. Main plant poisonings in livestock in the Nordic countries. In *Bioactive Compounds in Plants—Benefits and Risks for Man and Animals*; The Norwegian Academy of Science and Letters: Oslo, Norway, 2008; pp. 30–43.
2. Pain, S.; Revell, D.K. *Fodder Quality Specifications: Identifying Predictors of Preference between Hays*; Rural Industries Research and Development Corporation: Wagga Wagga, Australia, 2009.
3. Elliott, S.; Loudon, A. Effects of monoterpene odors on food selection by red deer calves (*Cervus elaphus*). *J. Chem. Ecol.* **1987**, *13*, 1343–1349. [CrossRef] [PubMed]
4. Estell, R.E.; Fredrickson, E.L.; Tellez, M.R.; Havstad, K.M.; Shupe, W.L.; Anderson, D.M.; Remmenga, M.D. Effects of volatile compounds on consumption of alfalfa pellets by sheep1. *J. Anim. Sci.* **1998**, *76*, 228–233. [CrossRef] [PubMed]
5. Forbes, J.M.; Kyriazakis, I. Food preferences in farm animals: Why don't they always choose wisely? *Proc. Nutr. Soc.* **1995**, *54*, 429. [CrossRef] [PubMed]
6. Van den Berg, M.; Lee, C.; Brown, W.Y.; Hinch, G.N. Does energy intake influence diet selection of novel forages by horses? *Livest. Sci.* **2016**, *186*, 6–15. [CrossRef]
7. Provenza, F.D. Postingestive feedback as an elementary determinant offood preference and intake in ruminants. *J. Range Manag.* **1995**, *48*, 2–17. [CrossRef]
8. Morand-Fehr, P. Dietary choices of goats at the trough. *Small Rumin. Res.* **2003**, *49*, 231–239. [CrossRef]
9. Gershenzon, J.; Dudareva, N. The function of terpene natural products in the natural world. *Nat. Chem. Biol.* **2007**, *3*, 408–414. [CrossRef]
10. Furstenburg, D.; van Hoven, W. Condensed tannin as anti-defoliate agent against browsing by giraffe (Giraffa camelopardalis) in the Kruger National Park. *Comp. Biochem. Physiol. Part A Physiol.* **1994**, *107*, 425–431. [CrossRef]
11. Vourc'h, G.; De Garine-Wichatitsky, M.; Labbé, A.; Rosolowski, D.; Martin, J.-L.; Fritz, H. Monoterpene Effect on Feeding Choice by Deer. *J. Chem. Ecol.* **2002**, *28*, 2411–2427. [CrossRef]
12. Richard, R.R.; Taylor, C.A., Jr.; Kothmann, M.M.; Huston, J.E. Volatile Oil Contents of Ashe and Redberry Juniper and its Relationship to Preference by Angora and Spanish Goats. *J. Range Manag.* **1996**, *49*, 35–41. [CrossRef]
13. Pilegaard, K.; Ladefoged, O. Wild plants that can cause poisoning in farm animals. *Dan. Veterinærtidsskrift* **1999**, *82*, 666–668.
14. Caloni, F.; Cortinovis, C. Plants poisonous to horses in Europe. *Equine Vet. Educ.* **2015**, *27*, 269–274. [CrossRef]
15. Wink, M. Mode of action and toxicology of plant toxins and poisonous plants. *Mitt. Jul. Kühn-Inst.* **2009**, *421*, 93–112.
16. Cortinovis, C.; Caloni, F. Alkaloid-Containing Plants Poisonous to Cattle and Horses in Europe. *Toxins* **2015**, *7*, 5301–5307. [CrossRef] [PubMed]
17. Anadón, A.; Martínez-Larrañaga, M.R.; Castellano, V. Chapter 78—Poisonous plants of Europe. In *Veterinary Toxicology*; Academic Press: Cambridge, MA, USA, 2012; pp. 1080–1094.

18. Humphreys, D.J. *Veterinary Toxicology*, 3rd ed.; Ballière Tindall: London, UK, 1988; p. 356.
19. Tribunskii, M.P. Colchicum autumnale Poisoning in Lambs. 1970. CAB Abstracts Archive. CAS Registry Number: 64-86-8. pp. 71–72. Available online: https://ezproxy.tiho-hannover.de/login?url=https://search.ebscohost.com/login.aspx?direct=true&db=ldh&AN=19712200888&lang=de&site=eds-live (accessed on 21 August 2023). (In Russian).
20. Debarnot, A. *L'intoxication par le Colchique d'automne*; L'Ecole Nationale Veterinaire D'Alfort: Paris, France, 1968.
21. Guitart, R.; Croubels, S.; Caloni, F.; Sachana, M.; Davanzo, F.; Vandenbroucke, V.; Berny, P. Animal poisoning in Europe. Part 1: Farm livestock and poultry. *Vet. J.* **2010**, *183*, 249–254. [CrossRef] [PubMed]
22. Caloni, F.; Scarpa, P.; Bruschera, L.; Pompa, V.; Davanzo, F. Communications. *J. Vet. Pharmacol. Ther.* **2003**, *26*, 183–184. [CrossRef]
23. Wagstaff, D.J. *International Poisonous Plants Checklist: An Evidence-Based Reference*; CRC Press: Boca Raton, FL, USA, 2008.
24. van Raamsdonk, L.W.D.; Ozinga, W.A.; Hoogenboom, L.A.P.; Mulder, P.P.J.; Mol, J.G.J.; Groot, M.J.; van der Fels-Klerx, H.J.; de Nijs, M. Exposure assessment of cattle via roughages to plants producing compounds of concern. *Food Chem.* **2015**, *189*, 27–37. [CrossRef]
25. Charlton, G.L.; Rutter, S.M. The behaviour of housed dairy cattle with and without pasture access A review. *Appl. Anim. Behav. Sci.* **2017**, *192*, 2–9. [CrossRef]
26. Kotowski, M.; Kotowska, D.; Biró, M.; Babai, D.; Sharifian, A.; Szentes, S.; Łuczaj, Ł.; Molnár, Z. Change in European Forage and Fodder Plant Indicator Sets over the Past 250 Years. *Rangel. Ecol. Manag.* **2023**, *88*, 159–173. [CrossRef]
27. Hegi, G. *Illustrierte Flora von Mitteleuropa*, 3rd ed.; J. F. Lehmann Verlag: München, Germany, 1981; Volume 3.
28. Fenwick, G. Bracken (*Pteridium aquilinum*)—Toxic effects and toxic constituents. *J. Sci. Food Agric.* **1989**, *46*, 147–173. [CrossRef]
29. Hubbs, J.C. Belladonna [Solanum] poisoning in pigs. *Vet. Med.* **1947**, *42*, 428–429. [PubMed]
30. Kanyuka, F.D. Pasture poisoning of pigs. *Veterinariya* **1954**, *31*, 56.
31. Priouzeau, M. Poisoning of cattle, pigs and horses by water hemlock (*Cicuta virosa*). *Recl. Med. Vet.* **1951**, *127*, 483–488.
32. Waret-Szkuta, A.; Jégou, L.; Lucas, M.N.; Gaide, N.; Morvan, H.; Martineau, G.-P. A case of eagle fern (*Pteridium aquilinum*) poisoning on a pig farm. *Porc. Health Manag.* **2021**, *7*, 2. [CrossRef] [PubMed]
33. Zentek, J.; Aboling, S.; Kamphues, J. Accident report: Animal nutrition in veterinary medicine—Actual cases: Houndstongue (*Cynoglossum officinale*) in pasture—A health hazard for horses. *Dtsch. Tierärztl. Wochenschr.* **1999**, *106*, 475–477. [PubMed]
34. Gardens, K.R.B. The International Plant Names Index and World Checklist of Vascular Plants 2023. Available online: https://powo.science.kew.org/ (accessed on 30 October 2023).
35. Zander, R. *Dictionary of Plant Names*; Ulmer: Stuttgart, Germany, 2002; p. 990.
36. Aboling, S.; (Institute of Animal Nutrition; University of Veterinary Medicine Hannover, Lower Saxony, Germany). Photodermatitis beim Pferd durch Pastinak auf der Weide. Personal communication, 2011.
37. Aboling, S.; (Institute of Animal Nutrition; University of Veterinary Medicine Hannover, Lower Saxony, Germany). Gesichtsschwellung bei einer Ziege durch Äsung von Brennessel. Personal communication, 2022.
38. Aboling, S.; (Institute of Animal Nutrition; University of Veterinary Medicine Hannover, Lower Saxony, Germany). Vergiftung durch Ahornkeimlinge. Personal communication, 2023.
39. Ramsbottom, J. Caroli Linnaei pan suecicus. *Trans. Bot. Soc. Edinb.* **1959**, *38*, 151–167. [CrossRef]
40. Winter, S.; Penker, M.; Kriechbaum, M. Integrating farmers' knowledge on toxic plants and grassland management: A case study on Colchicum autumnale in Austria. *Biodivers. Conserv.* **2011**, *20*, 1763–1787. [CrossRef]
41. Lange, G. Sicherung wirtschaftlicher Nutzung von Feuchtgrünlandstandorten unter Berücksichtigung der Sporenpflanze Sumpfschachtelhalm. Report on behalf of Deutsche Bundesstiftung Umwelt, Osnabrück. File number 26540–33/2. Unpublished work. 2012.
42. McNaughton, I.H.; Harper, J.L. Papaver L. *J. Ecol.* **1964**, *52*, 767–793. [CrossRef]
43. Pilkington, M. *Experimental Aftermath Grazing of Urban Neutral Grassland*; British Grasslands Society: Gateshead, UK, 2000; pp. 165–166.
44. Hellström, K.; Huhta, A.-P.; Rautio, P.; Tuomi, J.; Oksanen, J.; Laine, K. Use of sheep grazing in the restoration of semi-natural meadows in northern Finland. *Appl. Veg. Sci.* **2003**, *6*, 45–52. [CrossRef]
45. Turaj, S. A Surprising Weed, Yellow Rattle. University of New Hampshire Extension. 2010. Available online: http://www.fsa.usda.gov/Internet/FSA_File/stnr_nh_06172011_weed.pdf (accessed on 26 September 2023).
46. Dietl, W. Weeds of pastures and meadows in the European Alps. In *Biology and Ecology of Weeds*; Springer: Berlin/Heidelberg, Germany, 1982; pp. 375–385.
47. Engel, A.M.; El-Khatib, A.H.; Klevenhusen, F.; Weiss, M.; Aboling, S.; Sachse, B.; Schäfer, B.; Weigel, S.; Pieper, R.; Fischer-Tenhagen, C. Detection of Hypoglycin A and MCPrG Metabolites in the Milk and Urine of Pasture Dairy Cows after Intake of Sycamore Seedlings. *J. Agric. Food Chem.* **2013**, *71*, 10751–10760. [CrossRef]
48. Nielsen, C.; Ravn, H.P.; Nentwig, W.; Wade, M. *The Giant Hogweed Best Practice Manual*; Forest & Landscape Denmark: Hoersholm, Denmark, 2005; p. 44.
49. Vogt-Andersen, U.; Calov, B. Long-term effects of sheep grazing on giant hogweed (*Heracleum mantegazzianum*). *Hydrobiologia* **1996**, *340*, 277–284. [CrossRef]
50. Susanne Ohlsen, M.G.; Wohlsein, P.; Huckauf, A.; Lenzewski, N.; Aboling, S. Grazing Ecology of Sheep and Its Impact on Vegetation and Animal Health on Pastures Dominated by Common Ragwort (*Senecio jacobaea* L.)—Part 1: Vegetation. *Animals* **2022**, *22*, 1000. [CrossRef] [PubMed]

51. Ohlsen, M.G.S.; Wohlsein, P.; Huckauf, A.; Lenzewski, N.; Aboling, S. Grazing Ecology of Sheep and Its Impact on Vegetation and Animal Health on Pastures Dominated by Common Ragwort (*Senecio jacobaea* L.)—Part 2: Animal Health. *Animals* **2022**, *12*, 13 [CrossRef] [PubMed]
52. Angell, G.N. What! Sheep control tansy (Tanacetum)? *Ore. Fmr.* **1950**, *73*, 861–872.
53. Forenbacher, S. Schachtelhalmvergiftung der Pferde—Eine B_1-Avitaminose. *Schweiz. Arch. Tierheilkd.* **1952**, *94*, 153–171.
54. Aboling, S.; Scharmann, F.; Bunzel, D. Equine atypical myopathy: Selection of sycamore maple seedlings (*Acer pseudoplatanus* L.) in pasture horses is driven by seedling maturity and might be associated with phenolic compounds. *Vet. Rec.* **2020**, *187*, e116. [CrossRef] [PubMed]
55. Berendonk, C.; Cerff, D.; Hünting, K.; Wiedenfeld, H.; Becerra, J.; Kuschak, M. Pyrrolizidine alkaloid level in Senecio jacobaea and Senecio erraticus—*The* effect of plant organ and forage conservation. In *Grassland in a Changing World. Proceedings of the 23rd General Meeting of the European Grassland Federation, Kiel, Germany, 29 August–2 September 2010*; Mecke: Duderstadt, Germany, 2010.
56. van Galen, G.; Marcillaud Pitel, C.; Saegerman, C.; Patarin, F.; Amory, H.; Baily, J.D.; Cassart, D.; Gerber, V.; Hahn, C.; Harris, P.; et al. European outbreaks of atypical myopathy in grazing equids (2006–2009): Spatiotemporal distribution, history and clinical features. *Equine Vet. J.* **2012**, *44*, 614–620. [CrossRef]
57. Westermann, C.M.; Leeuwen, R.; Raamsdonk, L.W.D.; Mol, H.G.J. Hypoglycin A Concentrations in Maple Tree Species in the Netherlands and the Occurrence of Atypical Myopathy in Horses. *J. Vet. Intern. Med.* **2016**, *30*, 880. [CrossRef]
58. Amdal, S. Planteforgiftning pä sau i Kvaenangen i begynnelsen av mai 1989 [Plant poisoning in sheep in Kvaenangen at the beginning of May 1989]. *Nor. Veterinaertidsskrift* **1991**, *103*, 1031.
59. Pearce, O.D. Drought in the west (*Brassica campestis* and *Chenopodium album* poisoning in cattle). *Vet. Rec. J. Br. Vet. Assoc.* **1975**, *97*, 60.
60. Borg, P.J.V. Ecology of *Equisetum palustre* in Finland, with special reference to its role as a noxious weed. *Ann. Bot. Fenn.* **1971**, *8*, 93–141.
61. Malone, F.E.; Kennedy, S.; Reilly, G.A.C.; Woods, F.M. Bog asphodel (*Narthecium ossifragum*) poisoning in cattle. *Vet. Rec.* **1992**, *131*, 100–103. [CrossRef] [PubMed]
62. Wilson, A.L.; Johnston, W.G.; McCuska, H.B.; Bannatyne, C.C. Hemlock water droplot (*Oenanthe crocata*) poisoning in cattle. *Vet. Rec.* **1958**, *70*, 587–590.
63. Parker, W.H.; McCrea, C.T. Bracken (*Pteris aquilina*) poisoning of sheep in the North York Moors. *Vet. Rec.* **1965**, *77*, 861–866. [PubMed]
64. Craig, J.F.; Kerney, W.; Timoney, J.F. Ragwort Poisoning in Cattle and Cirrhosis of the Liver in Horses. *Vet. Rec.* **1930**, *10*, 159–174.
65. McLintock, J.; Fell, B.F. A case of acute ragwort poisoning in the horse. *Vet. Rec.* **1953**, *65*, 319–320.
66. Herweimjer, C.H.; Houter, L.F.D. Poisoning due to fat hen (*Chenopodium album*) in sheep. *Neth. J. Vet. Sci.* **1971**, *4*, 52–54.
67. Janowski, W.; Lewandowski, L. Water hemlock (*Cicuta virosa*) poisoning in cattle. *Med. Weter.* **1955**, *11*, 166–167.
68. Schulz, O.; Hommel, H. *Colchicum autumnale* poisoning in cattle. *Monatshefte Fur Vet.* **1975**, *30*, 333–334.
69. Greatorex, J.C. Some Unusual Cases of Plant Poisoning in Animals. *Vet. Rec.* **1966**, *78*, 725–727. [CrossRef]
70. Moyano, M.R.; Garcla, A.; Rueda, A.; Molina, A.M.; Mendez, A.; Infante, F. Echium vulgare and Senecio vulgaris Poisoning in Fighting Bulls. *J. Vet. Med. A* **2006**, *53*, 24–25. [CrossRef] [PubMed]
71. Ivens, P. Hogweed suspected of causing primary photosensitisation in a horse. *Vet. Rec.* **2011**, *169*, 81–82. [CrossRef] [PubMed]
72. Lang, E.; Sendil, C. *Hypericum crispum*, die Ursache einer bei Pflanzenfressern in der Türkei beobachteten Lichtkrankheit. *Tierärztliche Umsch.* **1971**, *26*, 73–76.
73. Schrader, A.; Schulz, O.; Völker, H.; Puls, H. Recent plant poisoning in ruminants of northern and eastern Germany. Communication from the practice for the practice. *Berl. Munch. Tierarztl. Wochenschr.* **2001**, *114*, 218–221. [PubMed]
74. Ford, E.J.H. A Preliminary Investigation of Photosensitization in Scottish Sheep. *J. Comp. Path.* **1964**, *74*, 37–45. [CrossRef] [PubMed]
75. Gunning, V.O. Suspected buttercup poisoning in a Jersey cow. *Br. Vet. J.* **1949**, *105*, 393. [CrossRef]
76. Heggstad, E. A fatal combination: Fat cattle in late pregnancy on poor pasture with buttercups. *Nor. Veterinærtidsskrift* **1989**, *101*, 935–936.
77. Piekarz, J. Buttercup poisoning in a horse. *Med. Weter.* **1981**, *37*, 658.
78. Braun, U.; Linggi, T.; Pospischil, A. Ultrasonographic findings in three cows with chronic ragwort (*Senecio alpinus*) poisoning. *Vet. Rec.* **1999**, *144*, 122–126. [CrossRef]
79. Grabner, A. Enzootic liver cirrhosis and the brain-liver syndrome in horses poisoned with *Senecio alpinus*. *Pferdeheilkunde* **1990**, *6*, 119–124. [CrossRef]
80. Vockrodt, H. Plant poisoning in farm animals (*Senecio jacobaea* and *Inula conyza*) and its prevention. *Monatshefte Vet.* **1973**, *28*, 59–62.
81. Simon, U. Goldhafer und Rinderkalzinose. *Tierzüchter* **1980**, *32*, 292–293.
82. Dirksen, G.; Plank, P.; Spieß, A.; Hänichen, T.; Dämmrich, K. Über eine enzootische "Kalzinose" beim Rind. *Dtsch. Tierärztliche Wochenschr.* **1970**, *77*, 321–338.
83. Wanner, M.; Kessler, J.; Martig, J.; Tontis, A. Enzootische Kalzinose bei Ziege und Rind in der Schweiz. *Schweiz. Arch. Tierheilk* **1986**, *128*, 151–160.

84. Flåøyen, A.; Bratberg, B.; Frøslie, A.; Grønstøl, H. Nephrotoxicity and hepatotoxicity in calves apparently caused by experimental feeding with *Narthecium ossifragum*. *Vet. Res. Commun.* **1995**, *19*, 63–73. [CrossRef] [PubMed]
85. Votion, D.M. Atypical myopathy: An update. *Pract.* **2016**, *38*, 241–246. [CrossRef]
86. Puschner, B.; Booth, M.C.; Tor, E.; Odermatt, A. Delphinium alkaloid toxicosis in cattle from Switzerland. In *Poisonous Plants and Related Toxins*; CABI Publishing: Wallingford, UK, 2004; pp. 38–43.
87. Aboling, S.; Rottmann, S.; Wolf, P.; Jahn-Falk, D.; Kamphues, J. Case Report: Complex Plant Poisoning in Heavily Pregnant Heifers in Germany. *J. Vet. Sci. Technol.* **2014**, *5*, 1. [CrossRef]
88. Guilhon, J. L'enseignement de la botanique et la mort subite au pré. *Bull. Acad. Vét. Fr.* **1988**, *61*, 267–271. [CrossRef]
89. König, K.; Delarocque, J.; Echelmeyer, J.; Eikelberg, D.; Schwarz, S.; Baumgärtner, W.; Feige, K.; Aboling, S. Field study on *Hypochaeris radicata* L. in horse pastures in Germany—Australian stringhalt as a climatic and ecological phenomenon. *Pferdeheilkunde Equine Med.* **2021**, *37*, 556–568. [CrossRef]
90. Evans, C.E. Bracken poisoning of farm animals. *Vet. Rec.* **1964**, *76*, 365–372.
91. Pohlenz, J.; Luthy, J.; Minder, H.; Bivetti, A. Enzootische Leberzirrhose beim Rind, verursacht durch Pyrrolizidinalkaloide nach Aufnahme von *Senecio alpinus* (Alpenkreuzkraut). *Schweiz. Arch. Fur Tierheilkd.* **1980**.
92. Giles, C.J. Outbreak of ragwort (*Senecio jacobaea*) poisoning in horses. *Equine Vet. J.* **1983**, *15*, 248–250. [CrossRef] [PubMed]
93. Miraglia, N.; Costantini, M.; Polidori, M.; Meineri, G.; Peiretti, P.G. Exploitation of a natural pasture by wild horses: Comparison between nutritive characteristics of the land and the nutrient requirements of the herds over a 2-year period. *Animal* **2008**, *2*, 410–418. [CrossRef] [PubMed]
94. Priymenko, N.; Domange, C. Two cases of sudden death in the horse. *Nouv. Prat. Vét. Équine* **2012**, 4+64–65.
95. Tatar, A.; Ottowski, F. Mass poisoning of cattle with cowbane (*Cicuta virosa* L.). *Zycie Wetarynaryjne* **1967**, *42*, 137–140.
96. Dijkstra, R.G.; Falkena, R. Cicutoxin poisoning in two ponies (*Cicuta virosa*, cowbane). *Tijdschr. Diergeneeskd.* **1980**, *106*, 1037–1039.
97. Panariti, E. Meadow saffron (*Colchicum autumnale*) intoxication in a nomadic Albanian sheep flock. *Vet. Hum. Toxicol.* **1996**, *38*, 227–228.
98. Simpson, D.J.; Wainwright, S.J.; Hipkin, C.R. Presence of 3-nitropropionic acid, in widely distributed pasture legumes in Britain. *Vet. Rec.* **1999**, *145*, 169–171. [CrossRef]
99. Nilsson, S.G. Mowing and grazing in late summer—Effects on herbs, especially orchids. *Sven. Bot. Tidskr.* **2018**, *112*, 171–175.
100. Barton, A.; Hewicker-Trautwein, M.; Ohnesorge, B. Leberzirrhose durch Jakobs-Kreuzkraut-Intoxikation. *Pferde Spieg.* **2010**, *13*, 36–39. [CrossRef]
101. Passemard, B.; Priymenko, N. Equine poisoning by Senecio in France. *Rev. Med. Vet.* **2007**, *158*, 425–430.
102. Brügmann, M.; Niemann, U.; Wiedenfeld, H.; Geburek, F.; Jünnemann, D.F. *Fallbericht aus der Pathologie: Tod eines Norwegers*; Niedersächsisches Landesamt für Verbraucherschutz und Lebensmittelsicherheit: Oldenburg, Germany, 2006.
103. Gunning, O.V. Poisoning in goats by black nightshade (*Solanum nigrum*). *Br. Vet. J.* **1949**, *105*, 473–474. [CrossRef]
104. Troxler, J.; Jans, F.; Floch, C. Utilization and conservation of dry marginal areas by grazing sheep and suckler cattle. II. Effect on the vegetation. *Rev. Suisse Agric.* **1990**, *22*, 231–238.
105. Fatyga, J. *Trends in Animal Production in Consecutive and Mixed Grazing Systems of Cattle and Sheep in the Sudeten*; Grassland Research Institute: Banská Bystrica, Slovakia, 1990; pp. 290–293.
106. Ludvíková, V.; Pavlů, V.V.; Gaisler, J.; Hejcman, M.; Pavlů, L. Long term defoliation by cattle grazing with and without trampling differently affects soil penetration resistance and plant species composition in *Agrostis capillaris* grassland. *Agric. Ecosyst. Environ.* **2014**, *197*, 204–211. [CrossRef]
107. Fiedler, A. Untersuchungen zum Tiergesundheitsstatus von Gallowayrindern auf Weiden mit Bewuchs von Jakobskreuzkraut (*Senecio jacobaea* L.) und zum möglichen Carry-over in essbare Gewebe. Research Thesis, University of Veterinary Medicine Hannover, Clinic for Cattle, Hannover, Germany, 2022.
108. Brumme, S. *Senecio jacobaea als Giftpflanze im Grünland: Relevanz für Tiergesundheit und Flächennutzung*; Hochschule Bremen: Bremen, Germany, 2015.
109. Vanselow, R.; Brendieck-Worm, C. Ground-ivy (*Glechoma hederacea*) and the cause of toxicosis in horses: In search of evidence. *Z. Ganzheitliche Tiermed.* **2012**, *26*, 88–93. [CrossRef]
110. Andrews, W.H. Some Recent Advances in our Knowledge of Plant Poisoning. *Proc. R. Soc. Med. Sect. Comp. Med.* **1926**, *19*, 2–13. [CrossRef]
111. Gupta, R.C. *Veterinary Toxicology*; Elsevier: Amsterdam, The Netherlands, 2007.
112. Roura, E.; Humphrey, B.; Tedó, G.; Ipharraguerre, I. Unfolding the codes of short-term feed appetence in farm and companion animals. A comparative oronasal nutrient sensing biology review. *Can. J. Anim. Sci.* **2008**, *88*, 535–558. [CrossRef]
113. Isselstein, J.; Jeangros, B.; Pavlu, V. Agronomic aspects of biodiversity targeted management of temperate grasslands in Europe—A review. *Agron. Res.* **2005**, *3*, 139–151.
114. Endenburg, N. Perceptions and attitudes towards horses in European societies. *Equine Vet. J.* **1999**, *31*, 38–41. [CrossRef]
115. Provenza, F.D. Behavioural mechanisms influencing use of plants with secondary metabolites. In *Herbivores—The Assessment of Intake, Digestibility and the Roles of Secondary Compounds*; Sandoval-Castro, C.A., Hovell, F.D., Torres-Acosta, J., Ayala-Burgos, A., Eds.; BSAS Publication Nottingham University Press: Nottingham, UK, 2004; Volume 34, pp. 183–195.
116. Provenza, F.D.; Pfister, J.A.; Cheney, C.D. Mechanisms of learning in diet selection with reference to phytotoxicosis in herbivores. *J. Range Manag.* **1992**, *45*, 36–45. [CrossRef]

117. Julian Taenzer, M.G.; Klevenhusen, F.; Saltzmann, J.; Dänicke, S.; These, A. Rumen Metabolism of Senecio Pyrrolizidine Alkaloids May Explain Why Cattle Tolerate Higher Doses than Monogasteric Species. *J. Agric. Food Chem.* **2022**, *70*, 10111–10120. [CrossRef] [PubMed]
118. Craig, A.M.; Latham, C.J.; Blythe, L.L.; Schmotzer, W.B.; O'Connor, O.A. Metabolism of toxic pyrrolizidine alkaloids from tansy ragwort (Senecio jacobaea) in ovine ruminal fluid under anaerobic conditions. *Appl. Environ. Microbiol.* **1992**, *58*, 2730–2736. [CrossRef] [PubMed]
119. Rahman, M.M.; Abdullah, R.B.; Wan Khadijah, W.E. A review of oxalate poisoning in domestic animals: Tolerance and performance aspects. *J. Anim. Physiol. Anim. Nutr.* **2012**, *97*, 605–614. [CrossRef] [PubMed]
120. Hünsche, A.K. *Untersuchungen zu möglichen Schadwirkungen einer Kontamination von Grundfutter mit getrocknetem Sumpfschachtelhalm (Equisetum palustre) bei Wiederkäuern und Ponys*; Bibliothek der Tierärztlichen Hochschule Hannover: Hanover, Germany, 2010.
121. Tribe, D.E. The behaviour of the grazing animal: A critical review of present knowledge. *Grass Forage Sci.* **1950**, *5*, 209–224. [CrossRef]
122. Fairley, R.A.; Williams, S.J.; Finnigan, C.E.; Greene, M.A.; Robinson, A.E.; Gill, J.M. The occurrence of hypocalcaemia in mid lactation dairy cattle after the consumption of large amounts of fat-hen (*Chenopodium album*). *N. Z. Vet. J.* **2012**, *60*, 261–262. [CrossRef] [PubMed]
123. Nicolau, A.; Barza, H.; Duca, H.; Creteanu, C.; May, H.; Popoviciu, A. Vergiftungen durch die Pflanze Glechoma hederacea beim Pferde. *Monatsh. Vet.* **1956**, *11*, 534–538.
124. Opitz von Boberfeld, W. *Grünlandlehre*; UTB: Stuttgart, Germany, 1994.
125. Musaev, F.A.; Zakharova, O.A.; James, L.F. Phytosociological Analysis of the Drained Landscape. *Volga Reg. Farml.* **2020**, *1*, 53–60. [CrossRef]
126. Widger, D. The Project Gutenberg eBook of The History of Don Quixote, by Miguel de Cervantes. Translated by J. Ormsby. Available online: https://www.gutenberg.org/files/5903/5903-h/5903-h.htm (accessed on 10 August 2023).
127. Leiber, F.; Walkenhorst, M.; Holinger, M. The relevance of feed diversity and choice in nutrition of ruminant livestock. *Landbauforschung* **2020**, *70*, 35. [CrossRef]
128. Voiton. Equine atypical myopathy: A review. *Vet. J.* **2008**, *178*, 185–190. [CrossRef]
129. Valberg, S.J.; Sponseller, B.T.; Hegeman, A.D. Seasonal pasture myopathy/atypical myopathy in North America associated with ingestion of hypoglycin A within seeds of the box elder tree. *Equine Vet. J.* **2013**, *45*, 419–426. [CrossRef]
130. Gessner, O. *Gift- und Arzneipflanzen von Mitteleuropa*; Carl Winter: Heidelberg, Bermany, 1974; p. 582.
131. Pieper, R.; Kroeger, S.; Weigend, M.; Hanschen, F.; Kroh, L.; Zentek, J. Hoary allyssum in hay: A "new" potential health hazard for horses. *Tierärztliche Prax. Ausg. Großtiere Nutztiere* **2010**, *38*, 171–176. [CrossRef]
132. Frohne, D.; Jensen, U. *Systematik des Pflanzenreichs unter besonderer Berücksichtigung chemischer Merkmale und pflanzlicher Drogen, 5. Auflage ed.*; Wissenschaftl. Verlagsgesellschaft: Stuttgart, Germany, 1998; p. 371.
133. Puls, R.; Newschwander, F.P.; Greenway, J.A. Cyanide Poisoning from Glyceria grandis S. Wats. ex Gray (Tall Mannagrass) in a British Columbia. *Can. Vet. J.* **1978**, *19*, 264–265. [PubMed]
134. Senguttuvan, J.; Paulsamy, S.; Karthika, K. Phytochemical analysis and evaluation of leaf and root parts of the medicinal herb, Hypochaeris radicata L. for in vitro antioxidant activities. *Asian Pac. J. Trop. Biomed.* **2014**, *4*, S359–S367. [CrossRef]
135. Stegelmeier, B.L.; Colegate, S.M.; Knoppel, E.L.; A., R.K.; Collert, M.G. Wild parsnip (*Pastinaca sativa*)-induced photosensitization. *Toxicon Off. J. Int. Soc. Toxinology* **2019**, *167*, 60–66. [CrossRef] [PubMed]
136. Ulvund, M.J. Chronic poisoning in a lamb grazing Phalaris arundinacea. *Acta Vet. Scand.* **1985**, *26*, 286. [CrossRef] [PubMed]
137. Stewart, J.; Liyou, O.; Wilson, G. Bighead in horses—Not an ancient disease. *Aust. Equine Vet.* **2010**, *29*, 55–62.
138. Frohne, D.; Pfänder, H.J. *Giftpflanzen. Ein Handbuch für Apotheker, Ärzte, Toxikologen und Biologen, 4. Auflage ed.*; Kiel: Stuttgart, Germany, 1999.
139. Wlaschitz, S.; Riedelberger, K.; Gruber, A. Goldhaferinduzierte enzootische Kalzinose bei Pferden in der österreichischen Dachstein-Tauern-Region. *Vet. Med. Austria* **2006**, *93*, 253–258.
140. Frncoglu, H.K.; Tate, M.; Ünal, S.; Dogruyol, L.; Özcan, I. A selection strategy for low toxin vetches (*Vicia sativa* spp.). *Turk. J. Agric. For.* **2007**, *31*, 303–311.

animals

MDPI

Article

Oak Acorn Poisoning in Cattle during Autumn 2022: A Case Series and Review of the Current Knowledge

Justine Eppe [1,*], Calixte Bayrou [1], Hélène Casalta [1], Dominique Cassart [2], Linde Gille [1], Margot Stipulanti [1], Jérôme Versyp [1] and Arnaud Sartelet [1]

1 Fundamental and Applied Research for Animals and Health Research Unit (FARAH), Clinical Department of Production Animals, Faculty of Veterinary Medicine, University of Liège, Quartier Vallée 2, Avenue de Cureghem 7A-7D, 4000 Liège, Belgium; calixte.bayrou@uliege.be (C.B.); hcasalta@uliege.be (H.C.); linde.gille@uliege.be (L.G.); margot.stipulanti@uliege.be (M.S.); jerome.versyp@uliege.be (J.V.); asartelet@uliege.be (A.S.)

2 Fundamental and Applied Research for Animals and Health Research Unit (FARAH), Department of Animal Pathology, Faculty of Veterinary Medicine, University of Liège, Quartier Vallée 2, Avenue de Cureghem 6, 4000 Liège, Belgium; dominique.cassart@uliege.be

* Correspondence: justine.eppe@uliege.be

Simple Summary: Oak is widespread in Europe and can cause poisoning in grazing animals. Cattle are particularly susceptible. Herein, we describe seven cattle from three different farms admitted to the clinic for ruminants of the University of Liège for suspected acorn poisoning during autumn 2022. The clinical signs were vague. Blood analysis indicated renal failure. Of the hospitalized animals, five out of the seven had to be euthanized due to relapse. Lesions observed at necropsy were mainly digestive erosions and ulcerations, oedema and renal hemorrhages. Histopathological examination revealed necrosis of the renal tubules. Acorn poisoning is a serious disease with no specific antidote and no characteristic symptoms. Animals tend to be identified as sick late, when renal failure is already established. Farmers should be made aware of the prevention of this disease, especially in years when acorns are abundant. Furthermore, there is no antidote for this intoxication.

Abstract: Oak poisoning is a known intoxication in grazing animals, but is slightly described in the literature. This case report describes 7 cattle from 3 different farms admitted to the clinic for ruminants of the University of Liège for suspected acorn poisoning in the autumn of 2022. The clinical signs were, anorexia, apathy with polyuria with low density. Further investigations led to the diagnosis of renal failure (blood urea 162 ± 88 mg/dL; blood creatinine 12 ± 4 mg/L). Supportive treatment, based on infusions (NaCl 0.9%) and electrolyte rebalancing, was administered and renal values were assessed every 24–48 h. Of these animals, 5/7 were euthanized. At necropsy, digestive erosions and ulcerations, oedema and renal hemorrhages, between the pyloric/caliceal cavity and the medulla were observed. Histopathological examination revealed necrosis of the renal tubules. The renal values of the two remaining animals were reduced, their general condition improved, and they were discharged. Acorn poisoning is a serious disease with no specific antidote or characteristic symptoms. Animals are identified as sick too late, when renal failure is already established. Farmers should be made more aware in order to prevent exposure, especially in years when acorns are abundant. Furthermore, there is no antidote for this intoxication.

Keywords: cow; intoxication; oak; case report; pasture; kidney tubular necrosis; *Quercus*; Toxicology

1. Introduction

Oak (*Quercus* spp.) is the most common tree species in northern Europe [1,2]. It grows both in forests and on the edge of meadows. Many species of oaks in north America and Europe are toxic [3]. Its buds, twigs, leaves and acorns are toxic for grazing animals [2,4–6].

Acorn poisoning in ruminants is caused by a direct irritant action of tannic acids on the gastrointestinal mucosa (rumenitis, ulcers, hemorrhages), and indirectly via the degradation products of tannic acid through rumen digestion in gallic acid and pyrogallol [4,7,8]. Sudden death in cases of massive ingestion may occur at this stage [2,4,9]. The absorption of gallic acid and pyrogallol causes—through a strong inflammation stimulus on the glomeruli and the proximal segment—a multifocal necrosis of the renal tubules with the creation of protein cylinders [3,7]. The precise mechanism of tannin compounds is still poorly understood, but it is speculated that they are able to plug the loop of Henle by entering the cell and forming a precipitate by combining with the cell membrane or nutrients, or that the simple irritant effect of tannins causes lesions in the organs where they accumulate most [2]. This leads to chronic renal failure several weeks after ingestion, with all its consequences: ascites, hydrothorax, perirenal, retroperitoneal and subcutaneous edema [4,7,9]. The lesions observed on the kidneys allow acorn poisoning to be diagnosed [3,7].

Most of the available literature on acorn poisoning in cattle has been set for a long time [6,7,10]. The clinical signs are loss of appetite, constipation followed by diarrhea, colic or emaciation [2,7,9]. The examined cattle show apathy. The heart rate may be within normal limits, but both tachycardia and bradycardia have been reported [6,7]. The respiratory rate may also be increased [6,7]. In the digestive system, inappetence and increased thirst are described. The rumen shows reduced motility or atonic, and the digestive sounds are weak or inaudible on auscultation [2,7]. Another observation is the frequent emission of a large quantity of very clear urine, with low specific gravity, containing glucose and albumin. The urinary pH is slightly acidic (mean = 6.7), certainly compared to the usual pH of 7–8 [7,11].

Blood biochemistry revealed uremia, creatinemia, dehydration and hypocalcemia [2,7,9]. Treatment generally consists of intensive supportive therapy: intravenous and subcutaneous infusions, correction of electrolyte disorders, glucose supplementation and charcoal administration. Renal parameters are monitored for a possible reversal of the renal failure [2,7]. This reversal is associated with an improvement in the animal's general condition, with a recovery of its appetite.

The prognosis of this intoxication is poor, with few (15–25%) affected animals recovering [7]. In view of the poor prognosis of this intoxication, the prevention and risk of acorns in a meadow must be emphasized and mitigated [7].

Although the oak is a tree commonly found in our grasslands, full case report descriptions, with supporting further laboratory analysis, are rare in the literature [4,9]. This case report provides a unique and contemporary description of seven cases of acorn poisoning in cattle received at the clinic for ruminants of the University of Liège (CRU Liège) in the autumn of 2022.

2. Cases Information

Between 13 October and 1 November 2022, the CRU Liège received seven young cattle, from three different farms, referred by the field veterinarians for suspected oak acorn poisoning. These animals were between 9 and 4 months old, all Belgian Blue Cattle breed (BBCB), with five females and two males.

All the animals presented at the CRU Liège had in common that they were grazed 3 weeks to 1 month before the onset of the clinical signs in meadows bordered by oak trees. The animals all showed apathy, anorexia, ruminal stasis and absence of feces. Some animals (4/7) also had an episode of hyperthermia. Their condition had not improved despite the treatments administered by the veterinarians (broad spectrum antibiotics, non-steroidal anti-inflammatories, vitamins A, E, zinc, selenium and magnet).

For two of the calves, blood analysis was performed by the referring veterinarian on a pooled blood sample, revealing renal insufficiency (urea = 341 mg/dL; normal range (No): 10–25 mg/dL; creatinine = 24.87 mg/dL; No: 0.4–1 mg/dL [12]), increased liver values (GLDH = 52.3 IU/L; No < 30 IU/L); Gamma GT (63 IU/L; No < 39 IU/L [12]) and hypocalcemia (67 mg/L; No = 83–104 mg/L [12]).

3. Clinical Findings

At the clinical examination, one calf arrived in sternal recumbency, while the others were standing (Table 1). All were lethargic. Some of the animals were hypothermic with a rectal temperature below 38.5 °C and one was hyperthermic (39.7 °C). Some of the animals showed signs of dehydration (capillary refill time greater than 2 s, skin folds greater than 3 s and enophthalmia). None of the animals had adenomegaly. At the cardiorespiratory auscultation, a heart murmur (systolic; 4 to 5/6 gradation) was found on two out of the seven cattle. One of these two cattle also had an arrhythmia. Abdominal auscultation revealed a decrease in digestive sounds. During the examination, the calves frequently urinated very clear urine.

Table 1. Clinical examination parameters observed on the 7 animals, compared with the normal range [12]. Values outside the norm are shown in bold.

Cattle	1	2	3	4	5	6	7	Reference Value
Sex	male	female	female	female	female	female	male	-
Weight (kg)	200	202	335	275	210	290	127	-
Age (month)	10	9	17	8	7	8	5	-
Respiratory rate (respiration/min)	**12**	24	**54**	**112**	40	32	24	20–40
Rectal temperature (°C)	**36.7**	**38.1**	**38**	**38.2**	**38.1**	38.9	**39.7**	38.5–39.5
Heart rate (Beats/min)	76	**52**	72	**108**	**108**	**110**	72	60–84
Capillary refill time (s)	**3–4**	<2	<2	<2	<2	<2	**3**	<2
Skin folds (s)	**5**	**3**	**3**	**3**	<2	<2	<2	<2
Enophtalmia (mm)	**3**	**2**	**2**	0	0	0	0	0
Mucous membranes	**Pale**	Pink	Pink	Pink	Pink	Pink	**Congestive**	Pink
Lymph nodes	Normal	Normal	Normal	Normal	Normal	Normal	Normal	Normal

4. Diagnostic Assessment

For two of the seven animals (n°5,7), renal failure was already confirmed before their arrival. In view of the history and clinical examinations, acorn poisoning remained as the first diagnostic hypothesis. Other differentials may be suspected, such as other intoxications or poisonings (plants with soluble oxalates, ethylene glycol, etc.) or any other causes of renal impairment (pyelonephritis, *Leptospira* spp., etc.).

Due to financial constraints and the poor prognosis, not all complementary examinations were carried out on all of the animals (Table 2). All of the cattle had a blood test for both urea and creatinine (plasma, Catalyst One®, Idexx, Westbroock, ME, USA). Blood gaz analysis (Vetstat®, Idexx, Westbroock, ME, USA was conducted on some of the animals (4/7)). Furthermore, L-lactate was determined on three of the animals (n°1, 2, 3) (Accutrend® Plus, Roche Diagnostics International AG, Rotkreuz, Switzerland). A complete hematology and biochemistry (biochemistry: Catalyst One®; hematology: Procyte Dx®, Idexx, Westbroock, ME, USA) was performed on three of the seven animals.

Table 2. Results of blood tests performed on cattle, compared to reference standards [12–14]. Values outside the norm are shown in bold. (-) indicates that the test has not been carried out for this animal.

Cattle	1	2	3	4	5	6	7	Reference Value
Urea (mg/dL)	**>130**	**>130**	**321**	**>130**	**>130**	**>130**	**124**	10–20
Creatinine (mg/dL)	**>13.6**	**>13.6**	**3.63**	**>13.6**	**>13.6**	**>13.6**	**>13.6**	0.4–1
Lactate (mmol/L)	**2.6**	1.3	**4**	-	-	-	-	<2
Blood pH	**6.96**	**7.24**	**7.22**	-	-	-	7.35	7.35–7.50

Table 2. *Cont.*

Cattle	1	2	3	4	5	6	7	Reference Value
HCO3− (mmol/L)	**10.6**	**17.7**	**13.6**	-	-	-	25.3	24–34
Na (mmol/L)	**<100**	**122**	**129**	-	-	-	**124**	134–14[illegible]
K+ (mmol/L)	**8.4**	1.9	2.3	-	-	-	1.4	3.9–5.[illegible]
Cl− (mmol/L)	**73**	**89**	**94**	-	-	-	**85**	94–105
Hematocrit (%)	29.5	27.1	23.6	-	-	-	-	22–33%
Hemoglobin (g/dL)	10.3	9.8	9	-	-	-	-	8–15
Leucocytes ($\times 10^9$/L)	**15.82**	**15.19**	10.05	-	-	-	-	4–12
Neutrophiles ($\times 10^9$/L)	**8.54**	**11.49**	**6.4**	-	-	-	-	0.6–4
Lymphocytes ($\times 10^9$/L)	6.54	2.85	2.64	-	-	-	-	2.5–7.5
Monocytes ($\times 10^9$/L)	0.63	0.75	0.94	-	-	-	-	0.025–0.[illegible]4
Eosinophiles ($\times 10^9$/L)	0.09	0.04	0.03	-	-	-	-	0.0–0.2[illegible]
Basophiles ($\times 10^9$/L)	**0.02**	**0.04**	**0.03**	-	-	-	-	0.00–0.[illegible]2
Platelets (K/µL)	508	579	426	-	-	-	-	100–80[illegible]
P (mg/L)	**>161**	**157**	**154**	-	-	-	-	42–77
Ca (mg/L)	**48**	**49**	**48**	-	-	-	-	83–10[illegible]
Total proteins (g/L)	68	77	**91**	-	-	-	-	70–85
Albumin (g/L)	35	40	42	-	-	-	-	32–42
Gamma-glutamyltransferase (IU/L)	28	26	38	-	-	-	-	11–39
Cholesterol (g/L)	0.23	0.81	0.98	-	-	-	-	0.73–2.[illegible]

The animal (n°7) presenting with arrhythmia received additional blood gaz and ion analysis to evaluate its potassium level. The potassium level was within the normal range for this calf.

The obtained results for the additional animals revealed hyperkalemia in one out of the four, hyponatremia and hypochloremia in all four and a reduced bicarbonate concentration and metabolic acidosis in three out of the four.

In terms of haematology and biochemistry, all seven animals showed severe uremia and creatinemia, three-thirds had neutrophilia, basophilia, hyperphosphataemia and hypocalcaemia, two-thirds had leukocytosis and hyperlactatemia and one-third had monocytosis.

Urinary analysis using a dip stick (Combur 10 Test, Roche Diagnostics Gmbh, Deutschland) was performed on one of the animals (n°3), which revealed glycosuria, acidic pH at 6 and proteinuria. The urine density was 1010 g/mL.

The combination of the above-described analysis led to a diagnosis of severe renal failure, with more or less electrolytic repercussions. The most likely diagnosis was acorn poisoning. Unfortunately, there is no specific diagnostic test for this disease. In view of the very severe kidney damage and the suspected diagnosis, the prognosis for the animals was poor.

5. Therapeutic Intervention

The basis of the treatment administered was a 0.9% NaCl intravenous therapy at a rate of 10 mL/kg/h. If oedema developed, the rate of infusion was reduced to 5 mL/kg/h and intravenous diuretics (Furosemide, 1 mg/kg, Dimazon®, Intervet International, MSD Animal, Bruxelles, Belgium) were added to the treatment. Depending on the treatment previously administered by the referring veterinarians and the analysis carried out on the animals, some animals received additional oral rehydration (DrenchDig®, Savetis, Quévert, France), rumen digestion stimulants (Rumiphyt®, Savetis, Quévert, France) calcium (Parpumag 30%®, Dechra, Northwick, United Kingdoms) and/or potassium chloride. Calcium and potassium chloride were distributed based on the ionic losses in the blood analysis results.

6. Follow Up

The urea and creatinine were measured every 24 to 48 h depending on the case. If the urea and creatinine levels did not reduce to the measurable range, which is associated with a non-improvement of the general condition (anorexia, persistent depression despite treatment) or even a worsening via the appearance of subcutaneous oedema, euthanasia of the animals was recommended.

During hospitalization, five of the seven cattle rapidly developed sub-ventral and pulmonary oedema (12 h after infusion) (Figure 1). These animals showed tachycardia, tachypnea and crackles on pulmonary auscultation. The infusion was decreased to 5 mL/kg/h and intravenous diuretics were added to the treatment of these animals.

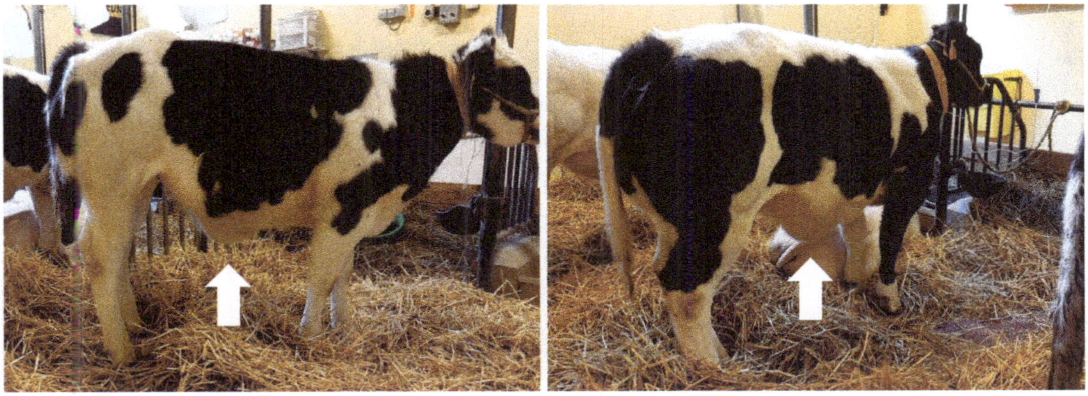

Figure 1. Sub-ventral oedema (arrow) of two cattle (n° 5 and 6).

For four of the seven cattle, no improvement, deterioration and/or unchanged renal parameters led to the decision to euthanize them 48 h after arrival.

Appetite returned in three of the seven cattle (n° 3, 4, 7) after 12 h of infusion, and the kidney values decreased after 24–48 h in two of these animals (cattle n° 3, 7). Within these three animals (cattle n° 3, 4, 7), two did not show sub-ventral edema. The treatments were continued for these animals while waiting for the renal parameters to return to a normal value (follow-up in Table 3). For cattle n°4, the renal values never decreased, and the sub-ventral oedemas persisted. He became anorexic again on the 5th day of hospitalization and was euthanized.

Table 3. Follow up renal values, urea (U) and creatinine (C) in mg/dL of cattle.

n°	DAY 1		DAY2		DAY3		DAY5		DAY10		DAY16	
	U	C	U	C	U	C	U	C	U	C	U	C
1	>130	>13.6										
2	>130	>13.6										
3	321	3.63	18	1.7					29	1	10	5.2
4	>130	>13.6	>130	>13.6	>130	>13.6	>130	>13.6				
5	>130	>13.6	>130	>13.6								
6	>130	>13.6	>130	>13.6								
7	124	>13.6	80	11.1	59	7	22	3.9	6	2.8		

7. Necropsy Findings

The five euthanized calves were necropsied. They were all in good shape. All had more or less severe generalized oedemas: subcutaneous cavitary oedema, hydrothorax, hydroperitoneum, hydropericardium. Some cavitary oedemas contained ten liters of fluid.

Lesions of the digestive tract were visible in all of the animals with congestion of the oesophagus, ulcerative abomasitis or small haemorrhagic foci (1–2 mm diameter) in the small intestine wall. The ulcers observed on the abomasal mucosa measured 6–7 mm by 3–4 mm and were located on top of the abomasal folds. The intestinal contents were liquid.

The other lesion systematically observed was a haemorrhagic zone at the junction between the medulla and the calyces/pyelic cavities in the renal interstitial tissue (Figure 2). These lesions were severe and involved both kidneys.

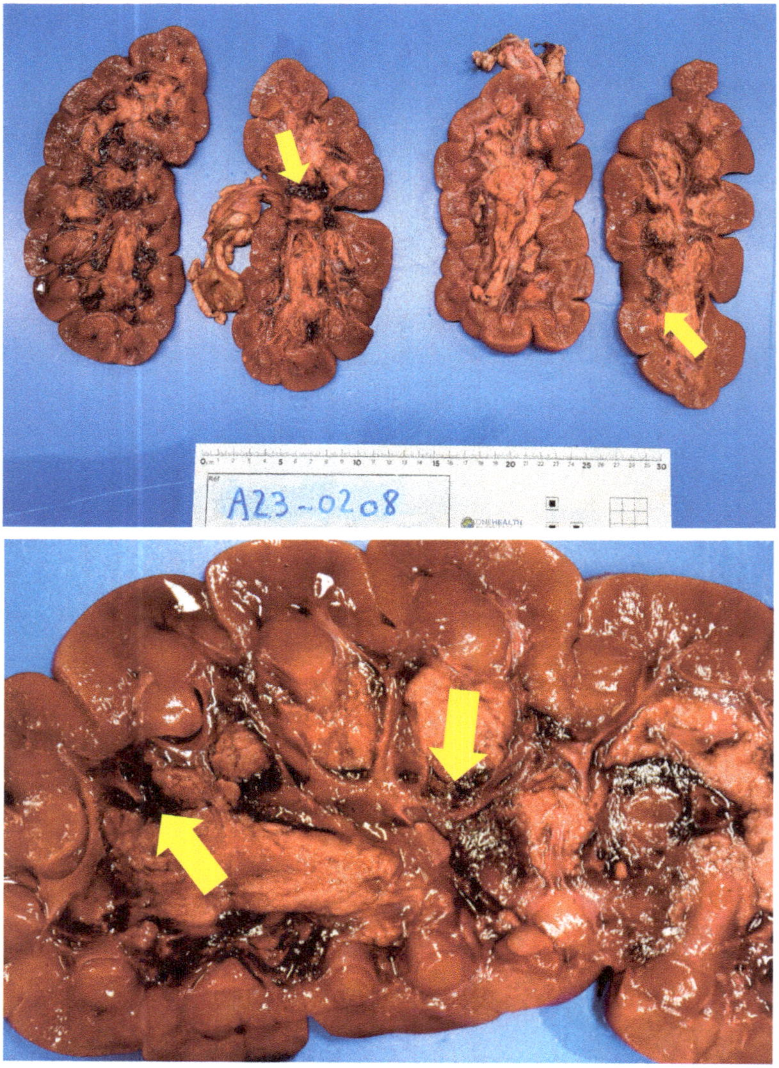

Figure 2. Kidneys of calf n° 6. Hemorrhagic areas can be seen at the junction between the medulla and the calyces/pyelic cavities (yellow arrows) as well as a pitted appearance of the cortex.

Microscopic examination of the kidneys revealed multifocal tubular necrosis with granular cylinders and foci of subacute interstitial nephritis (Figure 3). Microscopic lesions were concentrated in the renal tubules, with the glomeruli appearing intact. A different level of damage to the renal tubules was observed on the different histological slides performed on the animals, but this lesion was found in all of them. The tubules were dilated, their epithelium was absent and they were filled with necrotic debris.

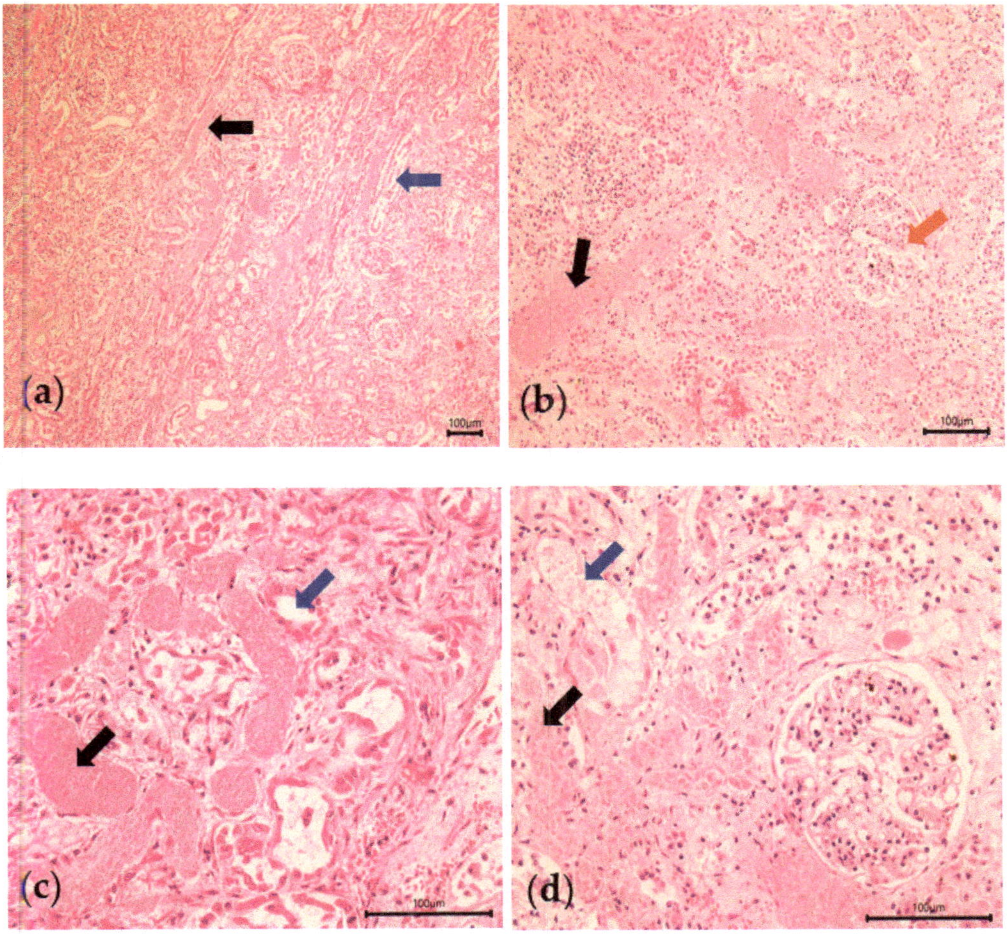

Figure 3. Histological images of kidney sections stained with hematoxylin-eosin at 100× (**a**), 200× (**b**), 400× (**c**,**d**) magnification. In (**a**), the extent of tubular necrosis and protein cylinders in the renal tubules can be seen. The black arrows show tubules containing hyaline necrosis, protein cylinders. The blue arrows show intact renal tubules. In (**b**) another image of hyaline tubular necrosis (black arrow) with a glomerulus (red arrow). In (**c**), section in necrotic tubules (black arrows), and in healthy tubules (blue arrows). In necrotic tubules, the tubular epithelium is no longer visible at all, and the lumen of the tubule is completely obstructed by necrotic material. In (**d**), another comparison between a healthy tubule (blue arrow) and a necrotic tubule (black arrow).

8. Outcome

Two cattle (n° 3 and 7) were discharged after several days of IV fluidotherapy (15 and 7 days, respectively). Both had their urea returned within the normal range, but their creatinine levels remained above the reference value (creatinine 5.2 and 2.8 mg/dL, respectively).

A follow-up 6 months later showed that the young cattle grew normally and were in good general health. For cattle n° 7, its fattening condition allowed it to be sold and sent to the slaughterhouse. There is no information on the outcome, nor its carcass weight. For cattle n° 3, a blood test was carried out. It revealed a normal uremia (140 mg/dlL), but the creatinine remains (2.2 mg/dL) slightly above the reference value. The heifer was stunted, with a chest circumference of 166 cm, the estimated weight being 370 kg (according to the formula [15]. Her clinical examination was within the normal range.

9. Discussion

The prevalence of exposure to acorns and subsequent poisoning depends on the year, the season, the age of the oak, the species, the maturity of the acorns and the weather conditions [3,6,7]. Summer drought, winds or windstorms are climatic risk factors for finding a lot of acorns on the meadow floor [3,8,16,17]. Further, it is reported that in some years, more acorns are produced by oaks [7]. Younger oaks also contain more gallic and tannic acids, with this concentration decreasing with the age of the oak [3]. Finally, some oak species are naturally more toxic, such as the English oak (*Quercus robur*) [3]. All of these considerations may explain why there is not the same frequency of acorn poisonings every year, although cases are reported every year at the Laboratory of Toxicology of the Faculty of Veterinary Medicine at Ghent University (LTGU) [17].

The toxic molecules contained in the leaves, acorns, blossoms and buds of the oak are gallic and tannic acids [3,4,8]. Oak leaves are most toxic in early spring, along with the buds and blossoms. The leaves tend to lose their toxicity with age, so that the brown leaves that fall in autumn are much less toxic. For acorns, the risk is greatest during the autumn period [6]. Acorn contains a high level of pyrogallic acids and are more toxic when immature and green. A mature acorn is considered to have little or no toxicity [3,7].

Acorn poisoning rarely affects grazing animals other than cattle, such as horses [3], goats, sheep [3] and camelids [18].

Not all cattle seem to be interested in acorns, and intoxication is often due to excessive curiosity on the part of young animals (0–3 years) [3,7,10], as was probably the case here, or the scarcity of grass in cases of drought [9,16]. Once the animals have tasted the acorns, buds or leaves, they may develop a craving for more [2]. Most of the affected animals were females (5/7). To the best of our knowledge, there is no information on the sex of animals poisoned by acorns. This sex ratio is probably due to the fact that males are principally fattened without access to the outside, whereas young females grow up in the meadows from April to November after one year of age.

In this case report, the episode followed a dry summer. The strength of this case report lays in the large amount of recent information collected on acorn poisoning cases, as well as the follow-up of these hospitalizations and the necropsy findings as little information is available in the current literature. Unfortunately, there is no clear protocol for the management of these clinical cases due to economical restraints, in some cases combined with the poor literature available, which leads to heterogeneity in data collection.

For chronic intoxication, clinical signs appear 1 to 3 weeks after grazing, which corresponds to what has previously been described [6,7]. Not very suggestive clinical signs, such as apathy, anorexia and an absence of feces, were reported. Loss of appetite is the first sign reported in the literature [6,7]. Constipation, a direct effect of tannic acid, followed by diarrhea with mucus and blood 1 to 8 days later, are also described [3,6]. The animals suffer kidney failure, which can be irreversible [19]. High water consumption and frequent emission of large amounts of urine have been reported [3,6]. In pregnant animals, abortion can also occur [2]. The animal stops gaining weight, stops growing, may be anemic and has declining subcutaneous oedema as a consequence of this renal failure [3,6]. In this case, the clinical signs related to renal failure were well observed, but not those related to the gastrointestinal tract, probably because the animals were already well advanced in the disease and anorexic. The hyperthermia that was also reported in four of the seven animals was not consistent with the symptomatology of acorn poisoning and was probably caused by another concomitant disorder.

The blood tests showed signs of renal failure: increased blood urea (7/7), creatinine (7/7), phosphate (3/3), potassium (1/4) and decreased blood chlorine (4/4) and calcium (3/3). Some of the animals were in metabolic acidosis (3/4). Dilution before the analysis was not systematically carried out when the values were above the measuring range in order to reduce the owner's costs. The decrease in ions and bicarbonate observed in some of the animals could be related to anorexia. Finally, some of the animals showed signs of inflammation on the haematology (leukocytosis: 2/3; neutrophilia: 3/3; monocytosis: 1/3). These results are consistent with what is described in the literature [7,20]

Tannic acids precipitate proteins on the cell membranes in the digestive tract, and cause erosion and ulcerations, in turn altering the absorption (injured mucosa) [5,19]. It can also increase vascular permeability [2] and edema, ascites, pleural or pericardial effusions may be observed, some containing several tens of liters of fluid [6]. The liver can also be affected [3,19]. The most characteristic lesion of this intoxication is the hemorrhage observed between the medulla and the calyces/pyelic cavity of the kidney. Microscopic examination allows a definite diagnosis to be made when tubular necrosis is observed [5,6]. The toxic compound causing these renal tubular necroses is gallic acid [19]. The suspected pathogenesis is a combination of tannin components with cell membrane proteins or nutrients, causing the cell to become necrotic, with these protein compounds then accumulating in the renal tubules [2]. All of the cattle examined at necropsy had these lesions. The microscopic examination provided a definitive diagnosis of these animals, revealing tubular necrosis. For cattle n°3, there is still doubt as to whether he suffered from acorn poisoning, as none of the animals in his herd have been autopsied and diagnosed, and there is no in vivo test to confirm acorn poisoning.

The prognosis is poor, with a fatal outcome in 75–85% of cases, mainly due to the non-reversibility of renal failure [7]. The number of acorns and the severity of renal failure probably play a role in the prognosis of this disease. One source indicates that eating 1 kg of green acorns for 15 days would be sufficient to induce intoxication in cattle [21]. The prognosis is poor with a urea blood level above 100 mg/dL, and hopeless above 300 mg/dL [2,21]. Regarding the poor prognosis of this poisoning, for economic and welfare reasons, it would be better to euthanize the animals and not attempt medical treatment. In this case report, two of the seven animals were discharged. The decrease in the renal values was already noticeable 24 h after the start of fluid therapy. Despite a satisfactory recovery, the surviving animals still retained their stunted growth, although they no longer relapsed at the renal level. A realistic protocol for the management of acorn poisoning could be to administer intravenous fluid therapy to the animals for 24 h, check their urea and creatinine, and if these reduce, continue fluid therapy until normal renal values are restored. For those animals whose renal values do not change, their case should be considered hopeless and euthanasia is recommended in these chronic cases.

There is no specific antidote for the tannins in oak, but several studies have compared diets containing salts (calcium hydroxide, hydrated lime) and/or vegetable oil to prevent this poisoning [6,8,16]. Trials with rations containing these additives are promising for limiting acorn poisoning, but this means supplementing animals that are on grassland with a feed that not all of them are willing to eat [6]. Once clinical signs are present, it is advisable to carry out an intensive supportive treatment based on intravenous or subcutaneous infusions and the correction of electrolyte disorders (calcium, potassium) [7]. A clinical case on a zebu (*Bos taurus indicus*) describes an effective treatment of acute renal failure caused by this intoxication with hemodialysis [20]. However, hemodialysis is too expensive to be used to treat production animals.

10. Conclusions

Exposure to acorns can have severe repercussions, which can cause serious after-effects several weeks after ingestion. The absence of an effective treatment reinforces the need to insist on prevention. Suspected cases are observed every year. Once clinical signs appear, it is often too late to save the animal. First, digestive lesions are present, followed by renal failure a few weeks after, with frequently irreversible lesions. Symptomatic treatment is the only solution. Uremia and creatininemia monitoring seems to be the solution for decision-making. However, this intoxication must be put into perspective, because it depends on many parameters, such as the type of oak, the climatic conditions and the year. As with all poisonous plants, meadows should be examined before the animal grazing period and should be avoided at the end of the season on pastures with a lot of acorns on the ground. Future studies should focus on the development of a diagnostic test, given the non-specific clinical signs and consequences of this disease.

Author Contributions: Acquisition of data: J.E., H.C., D.C., L.G., M.S., J.V. and A.S.; supervision: A.S.; writing-original draft preparation, J.E. and A.S.; writing—review and editing: C.B., H.C., D.C., L.G., M.S. and J.V. All authors have read and agreed to the published version of the manuscript.

Funding: This research received no external funding.

Institutional Review Board Statement: Not applicable.

Informed Consent Statement: Informed consent was obtained from all subjects involved in the study.

Data Availability Statement: All data are available in the manuscript.

Acknowledgments: The authors would like to thank the rural veterinarians and farmers or owners who participated in the data collection and everyone who contributed to the correction and the improvement of the manuscript.

Conflicts of Interest: The authors declare no conflict of interest.

References

1. Kremer, A.; Hipp, A.L. Oaks: An Evolutionary Success Story. *New Phytol.* **2020**, *226*, 987–1011. [CrossRef] [PubMed]
2. Bates, N. Autumn Plant Poisoning Hazards. *Livestock* **2019**, *24*, 239–245. [CrossRef]
3. Anadón, A.; Martínez-Larrañaga, M.R.; Castellano, V. Poisonous Plants of Europe. In *Veterinary Toxicology: Basic and Clinical Principles*; Elsevier: Amsterdam, The Netherlands, 2012; pp. 1080–1094. ISBN 9780123859266.
4. Plumlee, K.H. Chapter 25—Plants. In *Clinical Veterinary Toxicology*; Plumlee, K.H., Ed.; Mosby: Saint Louis, MO, USA, 2004; pp. 337–442. ISBN 978-0-323-01125-9.
5. Pérez, V.; Doce, R.R.; García-Pariente, C.; Hervás, G.; Carmen Ferreras, M.; Mantecón, Á.R.; Frutos, P. Oak Leaf (Quercus Pyrenaica) Poisoning in Cattle. *Res. Vet. Sci.* **2011**, *91*, 269–277. [CrossRef] [PubMed]
6. Dollahite, J.W.; Housholder, G.T.; Camp, B.J. *Oak Poisoning In Livestock*; Texas Agricultural Experiment Station Bulletin: College Station, TX, USA, 1966; pp. 1–8.
7. Stober, M.; Ziegler, H.P.; Von Benten, K. Acorn Poisoning in Cattle. *Bov. Pract.* **1976**, *11*, 36–41. [CrossRef]
8. Neser, J.A.; Boomker, J.; Cable, H. Oak (Quercus Rubor) Poisoning In Cattle. *J. S. Afr. Vet. Assoc.* **1982**, *53*, 151–155. [PubMed]
9. Frias, C.; Simões, P.B.A.; Cota, J.; Pissarra, H.; Nunes, T.P.; Hjerpe, C.A.; Lima, M.S.; Ecbhm, D. Case Report-Chronic Oak Toxicity (Quercus Suber) in Beef Cattle in the South of Portugal: 17 Cases (2014–2018). *Bov. Pract.* **2019**, *53*, 170–176. [CrossRef]
10. Marsh, C.D.; Clawson, A.B.; Marsh, H. *Oak-Leaf Poisoning of Domestic Animals*; Washington Government Printing Office: Washington, DC, USA, 1919; pp. 1–50.
11. Divers, T.J. Urinary Tract Diseases. In *Rebhun's Diseases of Dairy Cattle: Third Edition*; Elsevier: Amsterdam, The Netherlands, 2018; pp. 526–552. ISBN 9780323396622.
12. Divers, T.J.; Peek, S.F. The Clinical Examination. In *Rebhun's Diseases of Dairy Cattle: Third Edition*; Elsevier: Amsterdam, The Netherlands, 2018; pp. 2–16. ISBN 9780323396622.
13. Jones, M.L.; Allison, R.W. Evaluation of the Ruminant Complete Blood Cell Count. *Vet. Clin. North Am.-Food Anim. Pract.* **2007**, *23*, 377–402. [CrossRef] [PubMed]
14. Boulay, G.; Francoz, D.; Doré, E.; Dufour, S.; Veillette, M.; Badillo, M.; Bélanger, A.M.; Buczinski, S. Preoperative Cow-Side Lactatemia Measurement Predicts Negative Outcome in Holstein Dairy Cattle with Right Abomasal Disorders. *J. Dairy Sci.* **2014**, *97*, 212–221. [CrossRef] [PubMed]
15. Fastré, B. *Evaluation Du Poids et de l'état Corporel Chez Les Bovins Blanc-Bleu-Belge (BBB)*; ULiège: Liège, Belgium, 2012.
16. Vermeire, L.T.; Wester, D.B. Shinnery Oak Poisoning of Rangeland Cattle: Causes, Effects & Solutions. *Rangelands* **2001**, *23*, 19–21. [CrossRef]
17. Vandenbroucke, V.; Van Pelt, H.; De Backer, P.; Croubels, S. Animal Poisonings in Belgium: A Review of the Past Decade. *Vlaams Diergeneeskd. Tiijdschrift* **2010**, *79*, 259–268. [CrossRef]
18. Chamorro, M.F.; Passler, T.; Joiner, K.; Poppenga, R.H.; Bayne, J.; Walz, P.H. Case Report Rapport de Cas Acute Renal Failure in 2 Adult Llamas after Exposure to Oak Trees (*Quercus* Spp). *Can. Vet. J.* **2013**, *54*, 61–64. [PubMed]

19. Gwaltney-Brant, S.M. Renal Toxicity. In *Veterinary Toxicology: Basic and Clinical Principles*; Elsevier: Amsterdam, The Netherlands, 2012; pp. 264–277; ISBN 9780123859266.
20. Davy-Moyle, R.B.; Londoño, L.; Nelson, E.A.; Bandt, C. Treatment of Acute Kidney Injury Secondary to Oak Intoxication with Hemodialysis in a Miniature Zebu. *J. Vet. Emerg. Crit. Care* **2018**, *28*, 361–365. [CrossRef] [PubMed]
21. Boubet, B. L'intoxication Aux Glands. Une Pathologie Souvent Mortelle Chez Les Bovins. Available online: https://www.pleinchamp.com/actualite/l-intoxication-aux-glands-une-pathologie-souvent-mortelle-chez-les-bovins (accessed on 20 March 2023).

animals

MDPI

Article

Factors Affecting Toxic and Essential Trace Element Concentrations in Cow's Milk Produced in the State of Pernambuco, Brazil

Emanuel Felipe de Oliveira Filho [1,2], Marta López-Alonso [2], Guilherme Vieira Marcolino [3], Pierre Castro Soares [1], Carlos Herrero-Latorre [4], Carla Lopes de Mendonça [5], Nivaldo de Azevedo Costa [5] and Marta Miranda [6,*]

1 Department of Veterinary Medicine, Universidade Federal Rural de Pernambuco (UFRPE), Rua Dom Manoel de Medeiros, s/n, Dois Irmãos, Recife 52171-900, Brazil; felipe130188@gmail.com (E.F.d.O.F.); pcastro.pe@gmail.com (P.C.S.)

2 Department of Animal Pathology, Faculty of Veterinary, Campus Terra, University of Santiago de Compostela, 27002 Lugo, Spain; marta.lopez.alonso@usc.es

3 Health and Reproduction of Ruminants-UAG/UFRPE, Garanhuns 55292-270, Brazil; marcolinogv@outlook.com.br

4 Research Institute on Chemical and Biological Analysis, Analytical Chemistry, Nutrition and Bromatology Department, Faculty of Sciences, Campus Terra, University of Santiago de Compostela, 27002 Lugo, Spain; carlos.herrero@usc.es

5 Clinic of Cattle of Garanhuns/UFRPE, Campus Garanhuns, Av. Bom Pastor–Boa Vista, Garanhuns 55292-270, Brazil; carlalopes.mendonca@gmail.com (C.L.d.M.); na.costa@hotmail.com (N.d.A.C.)

6 Department of Anatomy, Animal Production and Clinical Veterinary Sciences, Faculty of Veterinary, University of Santiago de Compostela, 27002 Lugo, Spain

* Correspondence: marta.miranda@usc.es

Simple Summary: Milk is one of the main components of the human diet, mainly due to its mineral and protein content. But its contamination by heavy metals could produce a serious public health problem. In this study, we have determined the levels of toxic (Cd and Pb) and essential (Cu, Fe, and Zn) elements in raw milk from cows raised in the State of Pernambuco (Brazil). A high percentage of the samples had Pb levels above the tolerance limit established by Brazilian legislation, and the proximity of the farms to major roads is the main cause. Therefore, the consumption of milk produced under these conditions can be considered a risk to public health.

Abstract: The aim of this study was to provide information on the levels of toxic (Cd and Pb) and essential (Cu, Fe, and Zn) elements in cow's milk produced in the State of Pernambuco (Brazil). A total of 142 samples of raw milk were collected, and the concentrations of essential and toxic elements were determined using inductively coupled plasma-optical emission spectrometry. In almost 30% of the samples analyzed, the Pb content exceeded the maximum level established in the Brazilian legislation (0.05 mg/L). By contrast, in all the samples, the Cd content was below the maximum allowable level (0.02 mg/L). The essential trace elements Cu, Fe, and Zn were generally present at lower concentrations than reported in other studies and can be considered within the deficient range for cow's milk. Statistical and chemometric procedures were used to evaluate the main factors influencing the metal concentrations (proximity to major roads, presence of effluents, and milking method). The study findings demonstrate that the proximity of the farms to major roads influences the concentrations of Cd, Pb, and Cu and that this is the main factor explaining the Pb content of milk. In addition, the presence of effluents influenced the concentrations of Cu, while no relationship between the metal content and the milking method was observed. Thus, in accordance with the study findings, the consumption of cow's milk produced in the region can be considered a risk to public health due to the high concentrations of Pb and the low concentrations of other essential minerals such as Cu, Zn, and Fe in some of the milk samples.

Keywords: milk; cattle; toxic and essential trace elements; Pernambuco; Brazil

1. Introduction

Milk is a wholesome food and represents an important constituent of the human diet (especially for infants, schoolchildren, and the elderly) as it contains nutrients that are essential for growth, bone development, immune function, and other important physiological functions [1,2]. Milk is considered the most diverse natural food product in composition. In addition to being a good source of protein, fat, and carbohydrate, milk is an ideal source of macro- and microelements such as calcium (Ca), potassium (K), phosphorous (P), magnesium (Mg), zinc (Zn), iron (Fe), copper (Cu), manganese (Mn), and selenium (Se). However, milk can also contain toxic elements, the most important of which is lead (Pb), which is known to have deleterious effects on the developing nervous system of children [2,3]. Milk can also be contaminated by other toxic metals such as cadmium (Cd), mercury (Hg), arsenic (As), and nickel (Ni), and even by high concentrations of essential elements (such as Co, Cr, Cu, Fe, and Zn) [4,5].

The composition of milk is greatly influenced by the nutritional status of the cow, as well as the stage of lactation, management, productive stage, genetics, and breed [6]. Moreover, the essential trace element profile of milk, particularly toxic element residues, is largely affected by the environment where the cows are raised [3,5,7]. Heavy metals mainly enter cow's milk through cattle feed and drinking water (as well as via the atmosphere). The feed and water can, in turn, be contaminated through the soil via sewage sludge used as fertilizer, artificial fertilizers, metals used in fungicidal agents, and other agricultural chemicals, and also via wastewater from various industries. The risk of milk becoming contaminated is particularly high in areas affected by anthropogenic pollution, such as smelting or mining areas and highly industrialized regions, allowing for the transfer of metal contamination to the atmosphere, soil, water, animal feed, animals and their products, and finally to humans [2,3,5,8–10]. In the case of Pb, milk can become contaminated when cows graze and drink water at roadsides. In addition, factors related to the manufacturing practices (particularly hygiene during milking) and possible contamination from the equipment during processing [4,5] can also increase the concentration of this toxic element in milk. In recent years, the contamination of milk by toxic elements (particularly Pb) has been indicated to be one of the most serious aspects of environmental pollution for human health, because milk is widely consumed, especially by children [3]. Numerous studies have therefore been conducted in order to monitor the presence of toxic elements in milk and related products (see table of previous studies), particularly in developing countries where information from national monitoring programmes is scarce and legislation aimed at environmental protection is sometimes less restrictive than in other more developed countries.

The concentrations of metals in milk sometimes exceed the maximum limits recommended. According to the European Commission [11] and Codex Alimentarius Commission, the maximum limit for Pb is 0.020 mg/L, but the Brazilian legislation (Decree n°. 55871/65) [12] establishes a higher limit of 0.05 mg/L. Previous studies carried out in different states of Brazil indicated that Pb concentrations in milk exceeded the maximum limits [13–15]. However, at present there are no data on the concentrations of toxic and trace elements in cow's milk produced in the semi-arid region of Pernambuco. The study of the transfer of Pb to the rock–soil–plant–milk systems is essential in view of the high natural levels of Pb in the rock and soil in this region [16–18]. The dairy basin in the state of Pernambuco plays an important role in the local economy and in supplying milk, since it is an important producing area where milk is treated industrially and is also sold at fairs and local markets [17]. The rearing system in most of the properties in the region is extensive or semi-intensive, and some of the Pb ingested by cows via the consumption of contaminated forage could be transferred to the milk [16–18].

The main objective of the present study was to provide information about the levels of toxic (Cd and Pb) and essential (Cu, Fe, and Zn) elements in cow's milk produced in the State of Pernambuco and to determine the main factors influencing the concentrations of these elements (e.g., proximity to road, presence of effluents, and the milking method) to evaluate whether the consumption of cow's milk produced in the region can be considered a public health risk.

2. Materials and Methods

This study was conducted according to the guidelines of the Declaration of Helsinki and approved by the Ethics Committee of The Universidade Federal Rural de Pernambuco (protocol code 23082.009185/2017-15, approved 15 April 2017).

2.1. Sample Collection

The milk samples analyzed in this study (n = 142) were produced by cows raised in extensive and semi-intensive dairy farm systems in the mesoregion of Agreste de Pernambuco (Garanhuns microregion) (Figure 1). A total of 14 farms were visited and the samples were collected between September and December 2018. The animals were raised in corrals or paddocks, divided according to the organization of the owner, and were fed with native pasture, forage palm, corn, and soybean, with water and mineralized salt ad libitum. The distance from the farm in which each sample was taken and the main roads were calculated. The proximity of the farms to major roads was considered close when the distance was less than 3 km and distant otherwise. The farms were in a radius of 1 to 12 km to the main roads. Five farms (40 samples) were close, and the remaining nine (102 samples) far. Regarding effluents, the influence of hydrographic features (large and small rivers, as well as springs and other sources such as wells and dams) on the farms and access to these sources by the cows was taken into account. Eight farms (66 samples) had influence of effluents and the remaining six (76 samples) did not. The milking method was classified as manual (79 samples), performed manually by a responsible worker, or automatic (63 samples), carried out by means of milking machinery. Raw milk samples (15 mL each) were obtained from cows in the early lactation stage (11 to 100 days postpartum). The first 3 jets of milk were discarded and the next 15 mL was collected. Samples were stored at 4 °C in 15 mL sterile plastic containers with a lid and sent to the Clinical Pathology Laboratory of the Garanhuns Cattle Clinic (Universidade Federal Rural de Pernambuco, UFRPE), where they were then stored at −20 °C until analysis.

2.2. Reagents and Standard Solutions

All solutions were prepared using ultrapure water of resistance 18 MΩ cm^{-1} produced by a Milli-Q purification system (Millipore Corp., Bedford, MA, USA). Stock standard solutions of the elements (1000 mg/L) were of ultrapure grade (ICP Multi element standard solution IV certiPUR). Nitric acid (69%) was obtained from Merck (Poole, UK).

2.3. Sample Analysis

Samples were subjected to acid digestion before analysis. The microwave-assisted digestion procedure was carried out at the Research Support Center (CENAPESQ) of the Federal Rural University of Pernambuco (UFRPE). Briefly, each sample of raw bovine milk (5 mL) was placed in a glass flask with 10 mL of HNO3 and digested in a microwave oven (model MarsXpress-CEM Technology Inside) for 28 min (step 1: 110 °C for 8 min, step 2: 170 °C during 10 min, and step 3: 170 °C for 10 min.) Digested samples were filtered through pyramid-folded filter paper (weight 80 g m^2, filtration rate 20–25 s) into a new sterile tube.

The concentrations of Cd, Pb, Cu, Fe, and Zn were determined using atomic emission spectrometry with inductively coupled plasma (ICP-OES) (Optima 7000 DV, PerkinElmer, Waltham, MA, USA) in the Soil Chemistry Laboratory (DEPA) (UFRPE). All samples were analyzed in duplicate, and the concentrations were expressed in mg/L.

An analytical quality control programme was applied throughout the study. Blank samples were run alongside the test samples and the values thus obtained were subtracted from the sample readings. The limits of detection (LOD) in the acid digest were calculated as three times the standard deviation of the reagent blanks: 0.012 (Cd), 0.036 (Pb), 0.14 (Cu),

0.85 (Fe), and 0.93 (Zn) μg/L. The elemental concentrations of all samples analyzed were above the respective LODs. To check the accuracy of the analytical method, multi-element standard solutions were used for calibration and run with the samples. The precision of the method was expressed as the analytical recovery, which in all cases was within an acceptable range (90 to 110%), with a relative standard deviation (RSD) <10%.

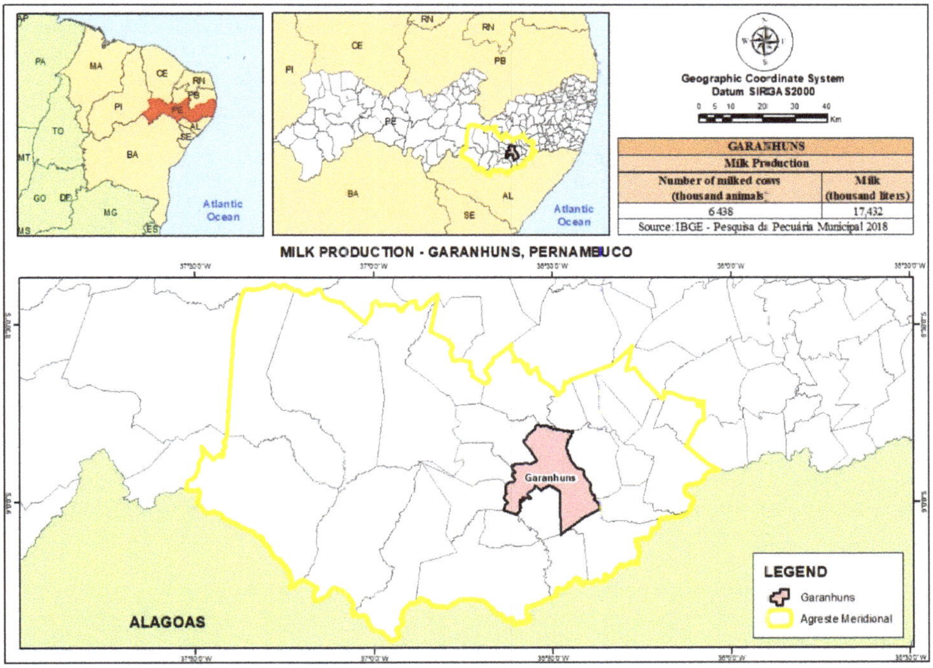

Figure 1. Municipality of Garanhuns and milk production in the Pernambuco dairy basin.

2.4. Statistical and Chemometric Analysis

An $X_{142\times5}$ matrix was used to analyze the data, with the rows corresponding to the 142 milk samples and the columns to the contents of the 5 toxic and essential metals determined using ICP-OES. Other information regarding the factors evaluated (proximity to main roads, presence of effluents, and milking system) were also included as qualitative variables in the data matrix. The data distribution was checked using the Kolmogorov–Smirnov test; as the data were not normally distributed, they were log-transformed before analysis and presented as geometric means. A general linear model (GLM) was used to evaluate the effect of the proximity of the farm to main roads (0: no; 1: yes); the presence of effluents (0: no; 1: yes); milking method (0: automatic, 1: manual); and their interactions in the toxic and essential trace element concentrations in milk. The statistical analyses were performed using IBM SPSS for Windows v.24 (IBM Corporation, Armonk, NY, USA) and test results were considered statistically significant at $p < 0.05$.

In addition, two unsupervised chemometric techniques, principal component analysis (PCA) and hierarchical cluster analysis (HCA), were used to reveal the latent structures residing in the data set and to evaluate the relationship between samples and variables. PCA was used to display the information contained in the data in a reduced dimension with minimum loss of data variance, and HCA (an unsupervised display chemometric technique, often used to complement PCA) was used to establish clusters of samples (and variables) based on the distance measures between them in the 5-multidimensional space [19]. All chemometric techniques were carried out using Statgraphics Centurion XVI v.16.1.15 (Statistical Graphics Corporation, Rockville, MD, USA).

3. Results and Discussion

3.1. Toxic and Essential Trace Element Concentrations in Milk

The toxic and essential trace element concentrations determined in the milk samples are presented in Table 1. The concentrations of these elements determined in other studies around the world are shown in Table 2 for comparative purposes.

Table 1. Toxic and essential trace element concentration (in mg/L) determined in milk samples in the present study.

	Mean ± SE	Median	GM	Range	25P	75P
Cd	0.007 ± 0.000	0.006	0.006	0.002–0.016	0.004	0.009
Pb	0.043 ± 0.002	0.040	0.040	0.013–0.098	0.028	0.053
Cu	0.020 ± 0.001	0.017	0.017	0.005–0.072	0.014	0.022
Fe	0.055 ± 0.004	0.041	0.040	0.002–0.283	0.026	0.073
Zn	0.621 ± 0.022	0.590	0.559	0.033–1.910	0.451	0.751

The mean Cd and Pb concentrations determined in the present study were 0.007 and 0.043 mg/L, respectively. The mean Cd concentrations were low and similar to those reported in other studies in unpolluted areas (generally below 0.010 mg/L) and much lower than those reported in previous studies in Brazil [14,20] and in polluted areas of developing countries in Asia, Africa, and South America (see Table 2). However, the concentrations of Pb are higher than those reported in other recent studies conducted in relatively unpolluted areas from Europe or North America (see Table 2), but lower than those reported in previous studies in Brazil [13–15,20]. The Pb concentrations determined in the present study are similar to those determined in some polluted regions of Iran [8] and Peru [21], but much lower than those reported in other polluted areas in Asia or Africa (see Table 2).

Considering the applicable legislation, the Pb content of 29.6% in the samples (42/142) was above the limit permitted under Brazilian law (Decree n. 55871/65), which is 0.05 mg/L [12], and 97% of the samples (138/142) exceeded the limit established in the European Union [11] and Codex Alimentarius Commission and the WHO, i.e., 0.020 mg/L. By contrast, the Cd content of all the samples was lower than 0.02 mg/L, the limit established by Brazilian law [12].

Previous studies conducted in Brasilia [22], Paraná [14], and Sao Luis [15] also reported Pb concentrations higher than the maximum limit. The Pb and Cd contents of milk depend on the proximity of polluted areas, crowded roads, the level of industrialization (see Table 2), and are also influenced by the control and legislation limits [3]. The concentrations determined in Western and Central Europe, the USA, and Canada are below levels considered to represent a risk, unlike in Brazil, Mexico, Peru, some parts of Asia or Africa, and polluted areas of Eastern Europe [3,23]. Recent studies have reported Pb and Cd concentrations in milk higher than 60 and 12 mg/kg, respectively, in some parts of India [3]; Pb concentrations in milk exceeding 13 mg/kg have been reported in Indonesia [24].

Considering the trace elements, the mean concentrations of Cu (0.020 mg/L), Fe (0.055 mg/L), and Zn (0.621 mg/L) were generally lower than those reported in other studies (see Table 2) and, in the case of Cu and Fe, can be considered within the deficient range for cow milk according to Puls [25] (deficient ranges are Cu: 0.010–0.020; Fe: <0.2; and Zn: <0.5 mg/L; adequate ranges are Cu: 0.05–0.6; Fe: 0.2–0.63; and Zn: 2.3–4 mg/L). These results indicate that milk produced in this region is not a good source of trace elements for the local population. The mean levels of Cu and Fe in raw cow's milk samples across the world ranged from 0.0136 to 36 mg/L and from 0.33 to 16.4 mg/L, respectively [3]. The presence of heavy metals such as Pb and Cd is associated with changes in the trace mineral profile of milk and negatively affects the nutritional quality of the product, e.g., by reducing the Fe content [2]. Trace element deficiencies in livestock are prevalent in different regions of Brazil [26], especially in several semi-arid climate areas. During the dry season, the pastures are generally overgrazed, leading to mineral deficiencies in the herds [16].

Table 2. Toxic and essential trace element concentrations (in mg/L) in cow's milk as determined in previous studies. Values are arithmetic means expressed in mg/kg or mg/L wet weight.

Country	Region	Cd	Pb	Cu	Fe	Zn	Reference
Algeria	Guelma area; polluted area	0.03	0.94	0.14	0.76	4.02	[27]
Algeria	Polluted area	0.030		0.239	1.43	5.98	[28]
Argentina	Rural areas Southeast of Córdoba		0.0023	0.0380	0.855	1.800	[29]
Bangladesh	Dairy Farms	0.024	0.015	0.064	0.333		[30]
	Small household	0.047	0.012	0.127	0.631		
Brazil	Paraná state; Pasteurized	0.018	0.281				[14]
	Paraná state; In Natura	0.031	0.181				
Brazil	State of Goiás (supermarkets)	0.05	0.24	0.49	0.96	3.73	[20]
Brazil	Industrial area	0.002	ND	0.063		3.87	[31]
	Non-industrial area	0.003	0.003	0.068		3.15	
Brazil	Vale of Paraíba region		0.23	1.73	1.05	4.59	[13]
China	Industrial	0.00015	0.00286				[32]
	Unpolluted	0.00013	0.00232				
China	Samples from Shandong and Shaanxi cities	0.00007	0.0014	0.0324	0.352	3.234	[33]
China	Ten main milk producing areas in China	0.00005	0.00175				[34]
Croatia	Four unpolluted areas	ND	ND–0.0071	0.06–0.07	0.26–0.30	3.7–4.8	[35]
Egypt	Beni Suef governorate	0.051	0.214	0.0953	8.994	6.29	[36]
England	Southern England; conventional farmland			0.0606	2.03	5.00	[37]
	Southern England; organic farmland			0.0524	0.66	4.51	
Ethiopia	Kosoye Amba-Rass, Tana-Abo, and Nara-Awdarda, North Gondar, Amhara Regional State	0.29	0.15	1.12		3.02	[38]
Hungary	Highway area	0.005	0.025	0.336	0.797	1.494	[39]
	Non-polluted	ND	0.012	0.137	0.788	2.241	
India	Mining areas		0.09–0.13	0.31–0.51	8.8–11.4	1.22–1.04	[40]
India	Ladakh, a trans-Himalayan high-altitude region	0.007–0.009	0.005–0.006	0.23–0.30	3.55–4.91	1.99–3.76	[41]
India	Industrial areas	0.02–0.07	0.05–0.20	0.07–0.35		1.22–20.94	[42]
India	Industrial area	0.096	0.480	0.090	3.97	6.09	[2]
	Non-industrial area	0.033	0.250	0.101	5.10	3.95	
Indonesia	City area, Padang	ND	13.6–20.6	1.17–2.17		28.8–53.1	[24]
Iran	Farms close to petroleum industries	0.0047	0.047				[8]
Iran	Arak city	0.00395	0.0125				[43]
Iran	Industrial regions of Iran	0.00111	0.0140	0.427		0.571	[44]
Iran	Lorestan province	0.10	2.72	0.14		3.07	[45]
Italy	Industrial area	ND	0.02	0.07–0.08	14.5–16 8	2.21–2.86	[46]
Italy	Calabria	0.0002	0.001	0.003		2.02	[1]
Kazakhstan	Almaty region; unpolluted	0.0027	0.0045				[47]
Korea	Supermarkets	0.00238	0.00335	0.3834		4.754	[48]
Kosovo	Rural areas	0.001	0.0017	0.018	0.426	3.151	[49]
Mexico	Areas irrigated with wastewater		0.03	0.01		0.71	[9]
Mexico	Puebla, industrial wastewater	0.002	0.024	0.030			[50]
Moscow	Moscow region	0.004–0.011	0.075–0.110	0.11–0.21	0.55–0.82	1.21–141	[51]
Pakistan	Sargodha; near traffic road	0.04–0.3	0.3–0.8				[52]
Peru	Near metallurgical complex	0.020	0.058				[21]
Peru	Mining-metallurgical industries	0.018	0.577				[10]
Poland	Lubuskie ProvinceOrganic farms	0.003–0.004	0.037–0.041	0.038–0.045	0.198–0.258	3.02–3.28	[53]
Poland	Low-level industrialization	<0.004	0.012	0.360		4.83	[54]
	High-level industrialization	<0.004	0.234	1.33		15.84	
Romania	Intermediate-level industrial	0.0039	0.120	2.4		4.8	[55]
	Intensive industrial	0.007	0.577	0.837		4.8	
	Small cattle farms	0.007	0.024	0.265		3.18	
Romania	No industry	0.006	0.066	0.30		2.5	[56]

Table 2. *Cont.*

Country	Region	Cd	Pb	Cu	Fe	Zn	Reference
Serbia	Novi Sad (Vojvodina) market	0.00349	0.0754	0.118			[57]
Spain	Organic farms	0.000135	0.000653	0.041	0.425	3.326	[58]
	Conventional farms	0.000098	0.000516	0.051	0.395	3.639	
	Conventional (supermarket)	0.000087	0.000267	0.069	0.351	3.933	
Spain	Unpolluted region; organic	ND	0.000519	0.039	0.271	2.851	[7]
	Unpolluted region; conventional	ND	0.000389	0.048	0.301	3.368	
Spain	Farms near mining and industrial area and highway traffic	<0.002	0.004				[59]
Sri Lanka	Four agro-climatic zones	0.001–0.002	0.005–0.02	0.02–0.12	0.49–3.15	1.49–2.93	[60]
Turkey	Local markets in the city of Edirne			0.138	3.1	3.4	[61]
Turkey	Iğdır City	0.0001–0.004	0.050	0.08–1.80		2.21–32.5	[62]
Turkey	Close to highways	0.39	1.85	0.62	4.2	1.85	[63]
Zambia	Farms near mining area		0.002				[64]

ND: not detected.

3.2. Effects of Proximity to Main Roads, Presence of Effluents, and Milking Method on Toxic and Trace Element Concentrations in Milk

A general linear model was applied to the data in order to evaluate the effect of factors that potentially influence the toxic and trace element concentrations in milk samples (proximity of the farm to major roads, presence of effluents in the vicinity of farm, and the milking method). The results are presented in Table 3, and it can be seen that the proximity to major roads (R) had a significant influence on the concentrations of Cd, Pb, and Cu in the milk samples. The presence of effluents in the vicinity of farms (E) only influenced the Cu concentration, while the milking method (M) did not influence the metallic profile of the samples.

Table 3. Summary of the general linear model used to evaluate the effect of proximity of major roads (R), the presence of effluents (E), milking method (M), and their interactions on the toxic and essential trace element concentration in milk in this study (* $p < 0.05$; ** $p < 0.01$).

Element	R	E	M	R × E	R × M	E × M	R × E × M
Cd	**	–	–	–	–	–	–
Pb	**	–	–	–	–	–	–
Cu	*	*	–	–	–	–	–
Fe	–	–	–	–	–	–	–
Zn	–	–	–	–	–	–	–

The proximity of the farm to major roads was the most important factor in the analysis, exerting a significant effect on the Cd, Pb, and Cu concentrations in milk, which were 94, 90, and 32% higher in the milk samples from farms close to major roads than in milk samples from farms distant from major roads (Figure 2). The influence of the proximity to major roads on toxic element accumulation in milk, blood, water, soils, forage, and other food products is well known [39,52,65]. Kodrik et al. [39] reported significantly higher levels of Cd (0.005 vs. ND mg/L), Pb (0.025 vs. 0.012 mg/L), and Cu (0.336 vs. 0.137 mg/L) in cow's milk originating from traffic-intensive areas in Hungary, whereas Fe concentrations (0.797 vs. 0.788 mg/L) were similar to those in milk produced in unpolluted green areas and the Zn concentrations were lower in the former than in the latter (1.494 vs. 2.241 mg/L). Tahir et al. [52] reported high levels of Cd (0.04–0.3 mg/L) and Pb (0.3–0.8 mg/L) in cow's milk in Pakistan, which were attributed to Cd- and Pb-contaminated feed, air pollution, and drinking water contaminated by dust from areas close to roads. Similar results were reported by Bigucu et al. [63] in areas of Turkey close to major roads. The levels of contamination in the milk were also higher in these studies than in the present study.

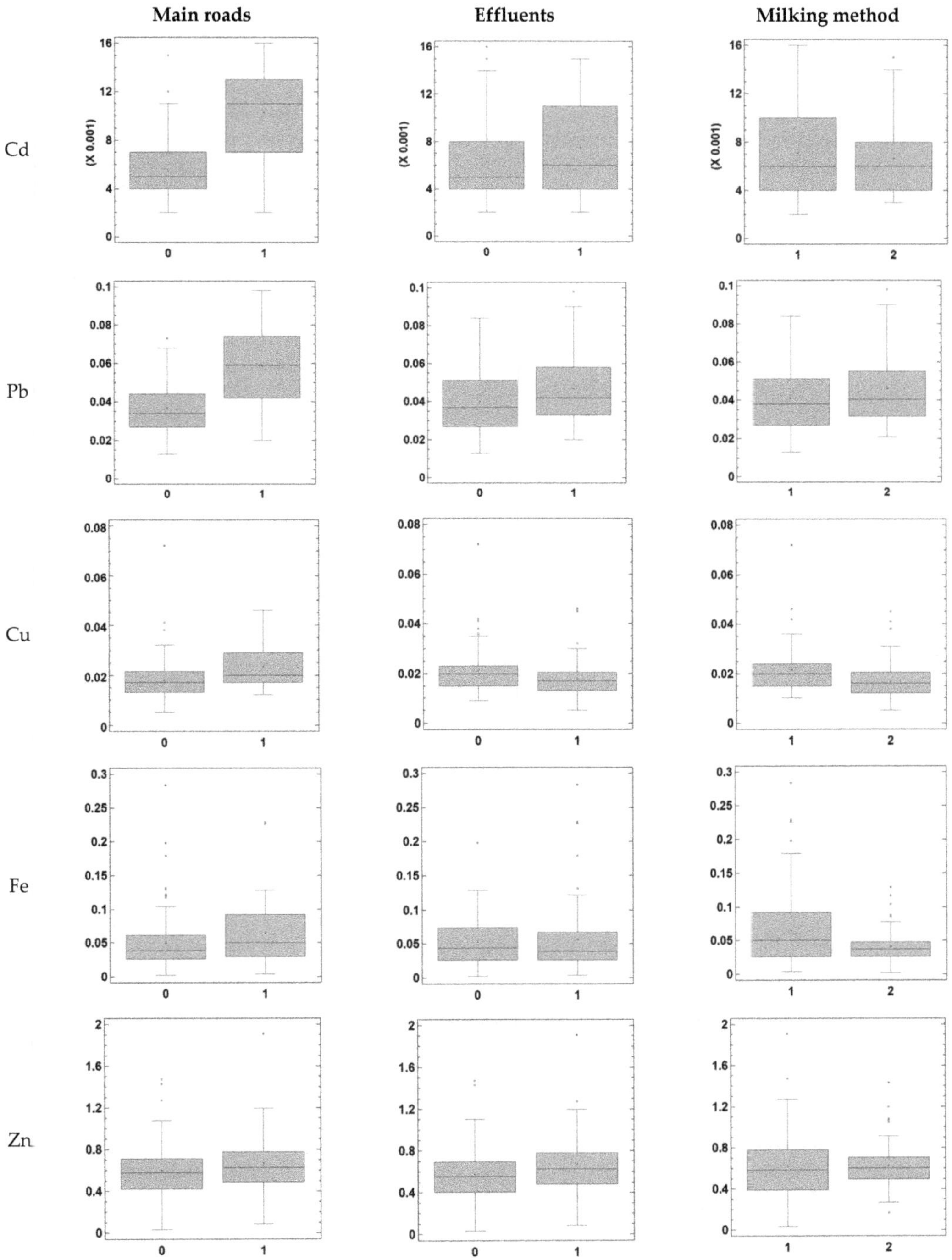

Figure 2. Box and whiskers plot showing the effect of proximity of farms to main roads, the presence of effluents, and milking method on toxic and trace element concentrations in milk (in mg/L).

Leaded petrol has caused more exposure to Pb than any other source worldwide, contaminating air, dust, soil, drinking water, and food crops, and it has caused harmfully high human blood Pb levels around the world, especially in children [66]. Lead persists in the environment and can bioaccumulate in bodies, with long-term effects lasting after the initial source of Pb has disappeared [66]. Some decades after Pb was banned in petrol, residues of this metal in milk have largely decreased, but emissions of toxic metals associated with motor vehicles are still considered among the most important sources of heavy metals in the environment, and high concentrations of heavy metals are found in soils near roads/highways [67]. Monitoring heavy metal concentrations in roadside environments therefore remains of great importance. The region studied, despite being a relatively unpolluted area without big industries, smelting, or mining areas, has high natural levels of Pb in the rock and soil [16], which could also contribute to the high levels of Pb in cow's milk in this region.

The presence of effluents was also found to be a significant factor regarding the Cu of the milk, although the samples from farms affected by effluents had lower Cu concentrations (23%) than samples from farms not affected by effluents. No significant interactions between the presence of effluents and the proximity to roads were found, and milk samples from farms close to and distant from main roads had lower Cu contents when they were close to sources of effluents. Concentrations of Cu in milk are known to be influenced by the environment, industrialization, and anthropogenic activities. A large part of the Pernambuco hydrographic basin, specifically the Garanhuns microregion, receives industrial effluents, and the water cannot be used to supply the city, as it contains toxic elements, and the rivers in these areas are contaminated and affected by anthropogenic activities [68]. As in the present study, Ogundiran et al. [69] reported that the Cu content was significantly higher in the milk from cows reared in an industrialized area than in milk from cows reared in unpolluted areas. These findings suggest that other factors/trace elements present in the effluents may interact negatively with the Cu metabolism in cows, leading to a significant reduction in the Cu concentration in the milk. For example, complex non-competitive negative interactions between Cu, sulphur, and molybdenum and competitive interactions with Cd and Zn for the metallothionein binding sites are known to occur in ruminants [6]. Other interactions between Cu and Cd [70,71] and Cu and Pb [72] have also been described in cattle. Increased levels of Cd and Pb interfere with the metabolism of the essential trace elements, in particular with the metabolism of Cu, with calves at risk of Cu deficiency [71]. Such interactions are sometimes difficult to interpret as they depend on the chemical form and relative concentrations of the elements in the environment and can involve more than two elements [2,72].

3.3. Chemometric Analysis: PCA and HCA

Chemometric analysis can be used for the detailed examination of large sets of data, enabling the visualization of complex interactions between samples and variables, and among samples and variables. Latent structures and relationships residing in the data matrix are commonly studied by means of two display chemometric techniques: PCA and HCA.

PCA was used for the primary evaluation of the 5-dimension data set. PCA transforms the autoscaled data matrix $X_{142\times5}$ into a product of two matrices: the score matrix S145xPC, which includes information about the samples, and the loadings matrix LPCxPC, which includes information related to the variables. When the number of principal components (PC) selected is smaller than the number of original variables, PCA produces a visualization of the data matrix in a reduced dimension, simplifies the original problem, and enables an examination of the relationships between samples and between variables through score and loadings plots, respectively [19].

In the present study, PCA was applied to the original autoscaled data matrix, and the first two PCs were considered sufficient to represent all the data. The first two principal components explained 75.38% of the total variance; the remaining PCs yielded eigenvalues <1, indicating poor information content (Table 4).

Table 4. Eigenvalues and variance explained by the principal components obtained from PCA.

PC	Eigenvalue	% Variance Explained	% Cumulative Variance Explained
1	2.416	48.33	48.33
2	1.352	27.05	75.38
3	0.778	15.56	90.94
4	0.392	7.84	98.78
5	0.060	1.22	100.00

This enabled the evaluation of the whole data set by using a 2D-score (or loading) plot, where the samples (or variables) are represented in the space defined by the first two principal components. An examination of the sample scores in this plot produced some interesting results. A natural separation of the samples into two groups according to the proximity of the farms to major roads was detected (Figure 3a), implying an evident influence of the traffic emissions on the metal content of the milk. Despite this "natural" separation, there was also some degree of overlap between the two categories in the 5-multidimensional space of the variables, as seen in the score-plot.

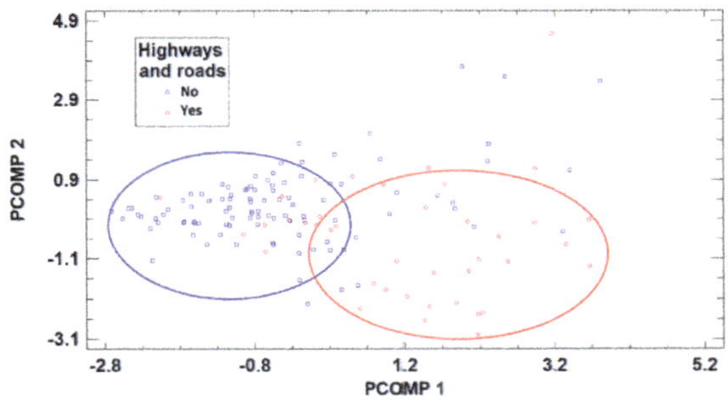

(a)

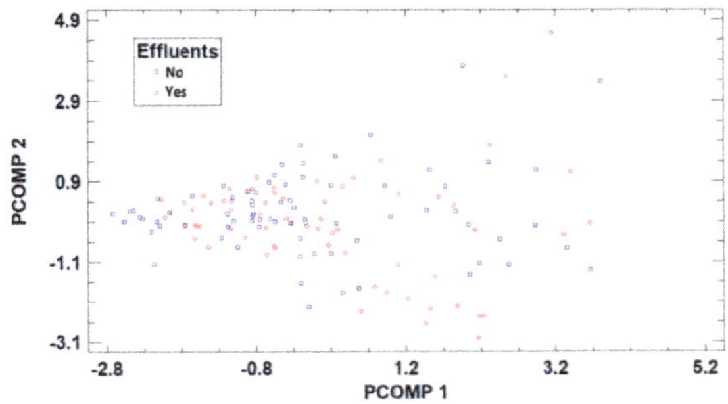

(b)

Figure 3. *Cont.*

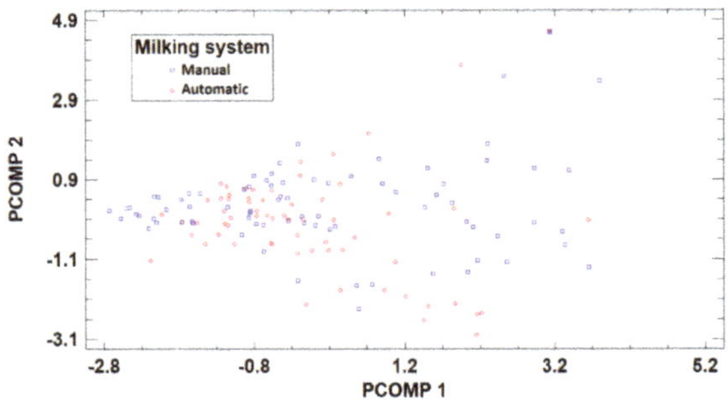

(c)

Figure 3. Twodimensional-score plot of the samples obtained using PCA according to (**a**) the proximity the farms to major roads, (**b**) the presence of effluents, and (**c**) the milking method.

The same separation in groups of samples was not detected for the other two factors studied, i.e., the presence of effluents in the vicinity of the farm (Figure 3b) and the milking method (Figure 3c). These results are consistent with previous findings established by using a GLM to evaluate the effect of the three factors considered, and they confirm that the proximity to major roads was the most significant factor affecting the concentration of metals in milk samples. The fact that the presence/absence of effluents in the proximity of the farms did not distinguish groups, which appeared randomly mixed thereby showing no significant influence of this factor on the metal content studied, reaffirms the possibility that other elements not determined in the study may interact negatively with Cu.

A PCA-biplot was constructed in order to study the relationship between variables in the whole data set. Both the samples and variables of the multivariate data were presented together (as scores and loadings) in the biplots. This type of plot reveals the relationships between the variables and also with the samples or groups of samples. In the case at hand, the 2D-biplot obtained, presented in Figure 4, showed the following: (i) a high degree of correlation between the three essential trace elements (Cu, Fe, and Zn), indicating a possible common origin, and (ii) an important association between toxic trace elements (Cd and Pb), also with a possible shared origin. The source in this case is probably from the emissions of motor vehicles from major roads, because the axis corresponding to these two variables coincided with the direction of the multidimensional space including the milk samples from farms close to major roads.

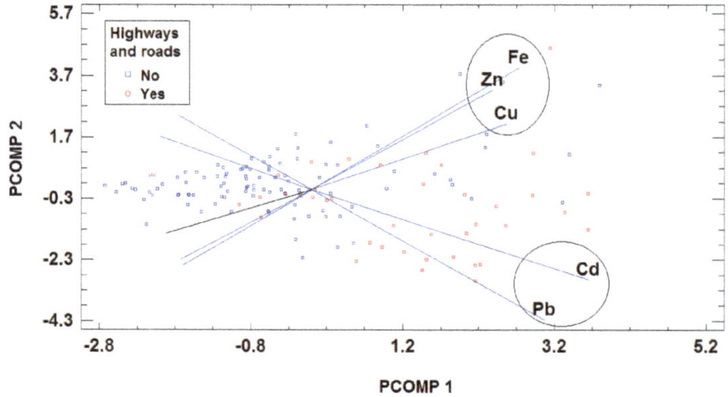

Figure 4. Biplot of the whole data set obtained using PCA. Samples are coded according to the proximity of major roads.

The second step of chemometric analysis consisted of using HCA. This unsupervised chemometric procedure searches for groups of samples (or variables) in the multidimensional space. It is based in an algorithm that arranges similar samples (or variables) into groups called clusters. The similarity between objects is calculated on the basis of the distance that separates them, considering that near samples in the 5-dimensional space of the variables will be very similar to each other. In the present case, the similarity was measured as the squared Euclidean distance, and the Ward method was used as an agglomerative algorithm procedure to identify clusters [73]. The dendrogram of the milk samples obtained when HCA was applied to the complete data set in the autoscaled $X_{142\times5}$ matrix can be seen in Figure 5. An examination of the dendrogram revealed the presence of two clusters of milk samples in the 5-dimensional space defined by the metal variables. The first cluster (A) is composed of samples from farms distant from major roads, while the second (B) mainly includes samples from farms close to major roads. The presence of samples near roads in cluster A and the appearance of samples far from roads in cluster B demonstrate some degree of overlap between clusters previously indicated in the PCA. This separation in two clear groups does not occur for either the milking type or the effluent factors, thus confirming the conclusion reached in the PCA that the most significant factor affecting the metal content of the samples is the traffic in the proximity of livestock farms. Moreover, the group of milk samples from farms far from major roads were more similar to each other than the group of milk samples from farms close to major roads were (Figure 5). This was an expected result as the distance of the farms from major roads was variable. It has been reported that the levels of heavy metal contamination (Pb and Cd are the most widely studied) in both soils and forage decreases to background levels with increasing distance on both sides of major roads [52,66]. Cattle and other animals that are grazed close to roads have been shown to have higher blood Pb levels than animals housed indoors [2,74], and the significant correlation between blood and milk Pb levels increases the potential of human exposure [2,74]. One of the most important causes of high Pb content in milk from rural, relatively unpolluted areas may be the proximity to major roads or transhumance along roads and/or motorways [63,75].

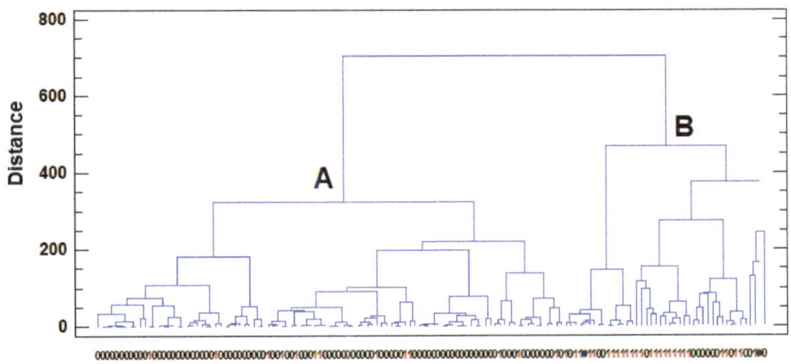

Figure 5. Dendrogram of the samples according to the proximity of major roads obtained using HCA (Squared Euclidean distance and Ward agglomerative method). (0 = far, 1 = close).

Additionally, HCA was used to evaluate associations between variables, and in this case, two clusters of variables were also detected (Figure 6): the first included the toxic metals, Cd and Pb, and the second comprised the essential trace elements. The variable arrangement may be directly linked to the differences in these metals according to the presence or absence of any major roads in the proximity of the farm where the milk samples were collected. In the plots, the metals are ordered from left to right on the basis of their capacity to distinguish between samples near and far from main roads. The variables in the first cluster (Cd and Pb) showed clearly different levels for samples on the basis of the road factor (with higher levels in milk from farms near main roads). Some differences in Cu concentrations were also observed for both classes, but with lower discriminatory capacity, while Fe and Zn showed similar or equal ranges. These results are consistent with those obtained using PCA and when the GLM was considered for testing the effect of the road factor in the metal content of milk samples.

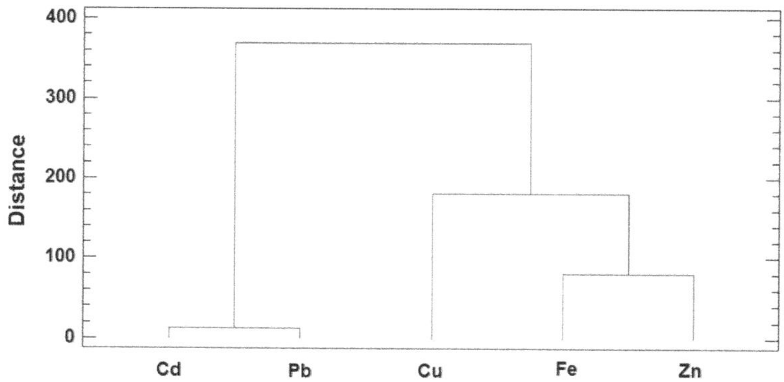

Figure 6. Dendrogram of the variables obtained by HCA (Squared Euclidean distance and Ward agglomerative method).

Taking into account the high concentrations of Pb in almost 30% of the milk samples analyzed, as well as the low content of other essential mineral elements, the consumption of milk from this region may represent a health risk to the population. More extensive studies that include larger sample sizes and different metals, and that determine Pb levels in the soil, pasture, feedstuffs and water should be conducted to clarify the risk associated with consumption of milk in the study region.

4. Conclusions

This study's findings indicate that raw cow's milk from the Pernambuco State contains high levels of Pb, but low levels of the essential elements Cu, Fe, and Zn. A high percentage of samples exceeded the maximum limit for Pb established in the current Brazilian legislation. This finding was attributed to the proximity of the farms to the major roads and was confirmed by using multivariate chemometric techniques (PCA and HCA). Both unsupervised chemometric approaches demonstrated the main influence of proximity to transportation infrastructures on the metal content of milk. On the other hand, neither the impact of the presence of effluents in the vicinity of the farm or the milking method had an important effect on the metal profile of the product. Thus, according to the metal levels detected, the consumption of cow's milk produced in this region can be considered a risk to public health due to the high levels of Pb and the low levels of essential minerals such as Cu, Zn, and Fe in some samples.

Author Contributions: Conceptualization, E.F.d.O.F., M.L.-A., G.V.M., P.C.S., C.L.d.M., N.d.A.C. and M.M.; methodology, E.F.d.O.F., M.L.-A., G.V.M., P.C.S. and M.M.; software, M.L.-A., C.H.-L. and M.M.; validation, E.F.d.O.F., M.L.-A., P.C.S. and M.M.; formal analysis, E.F.d.O.F., M.L.-A., P.C.S., C.H.-L. and M.M.; investigation, E.F.d.O.F., G.V.M., P.C.S., C.L.d.M. and N.d.A.C.; resources, M.L.-A., G.V.M., P.C.S., C.L.d.M., N.d.A.C. and M.M.; data curation, M.L.-A., C.H.-L. and M.M.; writing—original draft preparation, E.F.d.O.F., M.L.-A., G.V.M., P.C.S., C.H.-L. and M.M.; writing—review and editing, E.F.d.O.F., M.L.-A., C.H.-L. and M.M.; supervision, M.L.-A., P.C.S. and M.M. All authors have read and agreed to the published version of the manuscript.

Funding: This research received no external funding.

Institutional Review Board Statement: This study was conducted according to the guidelines of the Declaration of Helsinki, and approved by the Ethics Committee of The Universidade Federal Rural de Pernambuco (protocol code 23082.009185/2017-15, approved 15 April 2017).

Informed Consent Statement: Not applicable.

Data Availability Statement: Data sharing not applicable.

Conflicts of Interest: The authors declare no conflict of interest.

References

1. Licata, P.; Trombetta, D.; Cristani, M.; Giofrè, F.; Martino, D.; Calò, M.; Naccari, F. Levels of "Toxic" and "Essential" Metals in Samples of Bovine Milk from Various Dairy Farms in Calabria, Italy. *Environ. Int.* **2004**, *30*, 1–6. [CrossRef] [PubMed]
2. Patra, R.C.; Swarup, D.; Kumar, P.; Nandi, D.; Naresh, R.; Ali, S.L. Milk Trace Elements in Lactating Cows Environmentally Exposed to Higher Level of Lead and Cadmium around Different Industrial Units. *Sci. Total Environ.* **2008**, *404*, 36–43. [CrossRef]
3. Boudebbouz, A.; Boudalia, S.; Bousbia, A.; Habila, S.; Boussadia, M.I.; Gueroui, Y. Heavy Metals Levels in Raw Cow Milk and Health Risk Assessment across the Globe: A Systematic Review. *Sci. Total Environ.* **2021**, *751*, 141830. [CrossRef] [PubMed]
4. Enb, A.; Donia, M.A.; Rabou, N.S.A.; Abou-Arab, A.A.K.; El-Senaity, M.H. Chemical Composition of Raw Milk and Heavy Metals Behavior during Processing of Milk Products. *Glob. Vet.* **2009**, *3*, 268–275.
5. Pšenková, M.; Toman, R.; Tančin, V. Concentrations of Toxic Metals and Essential Elements in Raw Cow Milk from Areas with Potentially Undisturbed and Highly Disturbed Environment in Slovakia. *Environ. Sci. Pollut. Res.* **2020**, *27*, 26763–26772. [CrossRef] [PubMed]
6. Suttle, N.F. *Mineral Nutrition of Livestock*, 4th ed.; CABI: Wallingford, UK, 2010; ISBN 9781845934729.
7. Rodriguez-Bermúdez, R.; López-Alonso, M.; Miranda, M.; Fouz, R.; Orjales, I.; Herrero-Latorre, C. Chemometric Authentication of the Organic Status of Milk on the Basis of Trace Element Content. *Food Chem.* **2018**, *240*, 686–693. [CrossRef] [PubMed]
8. Norouzirad, R.; González-Montaña, J.-R.; Martínez-Pastor, F.; Hosseini, H.; Shahrouzian, A.; Khabazkhoob, M.; Ali Malayeri, F.; Moallem Bandani, H.; Paknejad, M.; Foroughi-nia, B.; et al. Lead and Cadmium Levels in Raw Bovine Milk and Dietary Risk Assessment in Areas near Petroleum Extraction Industries. *Sci. Total Environ.* **2018**, *635*, 308–314. [CrossRef]
9. Castro–González, N.P.; Calderón–Sánchez, F.; Pérez–Sato, M.; Soní–Guillermo, E.; Reyes–Cervantes, E. Health Risk Due to Chronic Heavy Metal Consumption via Cow's Milk Produced in Puebla, Mexico, in Irrigated Wastewater Areas. *Food Addit. Contam. Part. B Surveill.* **2019**, *12*, 38–44. [CrossRef]
10. Castro-Bedriñana, J.; Chirinos-Peinado, D.; Ríos-Ríos, E.; Machuca-Campuzano, M.; Gómez-Ventura, E. Dietary Risk of Milk-Contaminated with Lead and Cadmium in Areas near Mining-Metallurgical Industries in the Central Andes of Peru. *Ecotoxicol. Environ. Saf.* **2021**, *220*, 112382. [CrossRef]
11. Commision Regulation (EU). Setting Maximum Levels for Certain Contaminants in Foodstuffs; No 1881/2006. *Off. J. Eur. Union* **2006**, *364*, 5–24.
12. ANVISA Decree Number 55.871/65. 2016; pp. 1–23. Available online: http://anvisa.gov.br (accessed on 7 June 2023).
13. Soares, V.A.; Kus, M.M.M.; Peixoto, A.L.C.; Carrocci, J.S.; Salazar, R.F.S.; Izário Filho, H.J. Determination of Nutritional and Toxic Elements in Pasteurized Bovine Milk from Vale Do Paraiba Region (Brazil). *Food Control* **2010**, *21*, 45–49. [CrossRef]
14. Gomes, A.C.S.; Lindino, C.A.; Gonçalves Junior, A.C.; Gomes, G.D. Determinação de Cd, Cr e Pb No Leite e Na Alimentação Bovina Do Brasil. *Rev. Inst. Adolfo Lutz* **2013**, *72*, 211–218. [CrossRef]
15. Ribeiro Sant'Ana, M.A.; de Carvalho, T.C.; da Silva, I.F. Concentration of Heavy Metals in UHT Dairy Milk Available in the Markets of São Luís, Brazil, and Potential Health Risk to Children. *Food Chem.* **2021**, *346*, 128961. [CrossRef]
16. Silva, C.M.; Amaral, R.S.; Tabosa, J.N.; Santos Júnior, J.A.; Menezes, R.S.C.; Farias, E.E.G.; Bezerra, J.D.; Silvestre, R.G.; Oliveira, G.F. Pb-210 in Rock and Soils of the Semi-Arid Agreste Region of Pernambuco, Brazil. *Bull. Environ. Contam. Toxicol.* **2009**, *82*, 647–649. [CrossRef] [PubMed]
17. Silva, C.M.; Amaral, R.S.; Tabosa, J.N.; Júnior, J.A.S.; Menezes, R.S.C.; Ribeiro, F.C.A. Estimation of Dose Due to Ingestion of 210Pb in Milk from Dairy Cattle in the Semi-Arid Region of Pernambuco, Brazil. *Bull. Environ. Contam. Toxicol.* **2010**, *85*, 103–108. [CrossRef]
18. Silva, C.M.; Amaral, R.S.; Ribeiro, F.C.A.; Tabosa, J.N.; Júnior, J.A.S.; Menezes, R.S.C. Lead Poisoning Risk for Dairy Cows in the Semi-Arid Region of Pernambuco, Brazil. *Bull. Environ. Contam. Toxicol.* **2011**, *86*, 199–202. [CrossRef]
19. Massart, D.L. *Chemometrics: A Textbook*; Elsevier Science Publishers: Amsterdam, The Netherlands, 1988; Volume 2.
20. Gonçalves, J.; Mesquita, A.; Gonçalves, R. Determinação de metais pesados em leite integral bovino pasteurizado no estado de goiás determining heavy metals in pasteurized whole bovine milk in state of goiás. *Ciência Anim Bras.* **2008**, *9*, 365–374.
21. Chirinos-Peinado, D.M.; Castro-Bedriñana, J.I. Lead and Cadmium Blood Levels and Transfer to Milk in Cattle Reared in a Mining Area. *Heliyon* **2020**, *6*, e03579. [CrossRef]
22. De Castro, C.S.; Arruda, A.F.; Da Cunha, L.R.; SouzaDe, J.R.; Braga, J.W.; Dórea, J.G. Toxic Metals (Pb and Cd) and Their Respective Antagonists (Ca and Zn) in Infant Formulas and Milk Marketed in Brasilia, Brazil. *Int. J. Environ. Res. Public Health* **2010**, *7*, 4062–4077. [CrossRef]
23. Zwierzchowski, G.; Ametaj, B.N. Mineral Elements in the Raw Milk of Several Dairy Farms in the Province of Alberta. *Foods* **2019**, *8*, 345. [CrossRef]
24. Ketut Budaraga, I.; Salihat, R.A. Heavy Metals Analysis (Cd, Pb, Zn, Cu, Cr) and Calcium in Padang and Padang Panjang Fresh Cow's Milk. *IOP Conf. Ser. Earth Environ. Sci.* **2022**, *1038*, 012076. [CrossRef]

25. Puls, R. *Mineral Levels in Animal Health*, 2nd ed.; Clearbrook: Sherpa International, BC, Canada, 1994; ISBN 096934290X.
26. Bomjardim, H.A.; Oliveira, C.M.C.; Silveira, J.A.S.; Silva, N.S.; Duarte, M.D.; Faial, K.C.F.; Brito, M.F.; Barbosa, J.D. Deficiências Minerais Em Vacas Em Lactação Da Bacia Leiteira Do Município de Rondon Do Pará, Estado Do Pará. *Pesqui. Veterinária Bras.* **2015**, *35*, 409–416. [CrossRef]
27. Boudebbouz, A.; Boudalia, S.; Bousbia, A.; Gueroui, Y.; Boussadia, M.I.; Chelaghmia, M.L.; Zebsa, R.; Affoune, A.M.; Symeon, G.K. Determination of Heavy Metal Levels and Health Risk Assessment of Raw Cow Milk in Guelma Region, Algeria. *Biol. Trace Elem. Res.* **2023**, *201*, 1704–1716. [CrossRef] [PubMed]
28. Bousbia, A.; Boudalia, S.; Gueroui, Y.; Ghebache, R.; Amrouchi, M.; Belase, B.; Meguelati, S.; Belkheir, B.; Benidir, M.; Chelaghmia, M.L. Heavy metals concentrations in raw cow milk produced in the different livestock farming types in guelma province (algeria): Contamination and risk assessment of consumption. *J. Anim. Plant Sci.* **2019**, *29*, 386–395.
29. Pérez-Carrera, A.L.; Arellano, F.E.; Fernández-Cirelli, A. Concentration of Trace Elements in Raw Milk from Cows in the Southeast of Córdoba Province, Argentina. *Dairy. Sci. Technol.* **2016**, *96*, 591–602. [CrossRef]
30. Muhib, M.I.; Chowdhury, M.A.Z.; Easha, N.J.; Rahman, M.M.; Shammi, M.; Fardous, Z.; Bari, M.L.; Uddin, M.K.; Kurasaki, M.; Alam, M.K. Investigation of Heavy Metal Contents in Cow Milk Samples from Area of Dhaka, Bangladesh. *Int. J. Food Contam.* **2016**, *3*, 16. [CrossRef]
31. Souza, M.V.D.; Vianna, M.W.D.S.; Zandim, B.M.; Fernandes, R.B.A.; Fontes, M.P.F. Metais Pesados Em Amostras Biológicas de Bovinos. *Ciência Rural* **2009**, *39*, 1774–1781. [CrossRef]
32. Su, C.; Liu, H.; Qu, X.; Zhou, X.; Gao, Y.; Yang, H.; Zheng, N.; Wang, J. Heavy Metals in Raw Milk and Dietary Exposure Assessment in the Vicinity of Leather-Processing Plants. *Biol. Trace Elem. Res.* **2021**, *199*, 3303–3311. [CrossRef]
33. Zhou, X.; Qu, X.; Zhao, S.; Wang, J.; Li, S.; Zheng, N. Analysis of 22 Elements in Milk, Feed, and Water of Dairy Cow, Goat, and Buffalo from Different Regions of China. *Biol. Trace Elem. Res.* **2017**, *176*, 120–129. [CrossRef] [PubMed]
34. Zhou, X.; Zheng, N.; Su, C.; Wang, J.; Soyeurt, H. Relationships between Pb, As, Cr, and Cd in Individual Cows' Milk and Milk Composition and Heavy Metal Contents in Water, Silage, and Soil. *Environ. Pollut.* **2019**, *255*, 113322. [CrossRef]
35. Bilandžić, N.; Čalopek, B.; Sedak, M.; Đokić, M.; Gajger, I.T.; Murati, T.; Kmetič, I. Essential and Potentially Toxic Elements in Raw Milk from Different Geographical Regions of Croatia and Their Health Risk Assessment in the Adult Population. *J. Food Compos. Anal.* **2021**, *104*, 104152. [CrossRef]
36. Meshref, A.M.S.; Moselhy, W.A.; Hassan, N.E.-H.Y. Heavy Metals and Trace Elements Levels in Milk and Milk Products. *J. Food Meas. Charact.* **2014**, *8*, 381–388. [CrossRef]
37. Qin, N.; Faludi, G.; Beauclercq, S.; Pitt, J.; Desnica, N.; Pétursdóttir, Á.; Newton, E.E.; Angelidis, A.; Givens, I.; Juniper, D.; et al. Macromineral and Trace Element Concentrations and Their Seasonal Variation in Milk from Organic and Conventional Dairy Herds. *Food Chem.* **2021**, *359*, 129865. [CrossRef] [PubMed]
38. Akele, M.L.; Abebe, D.Z.; Alemu, A.K.; Assefa, A.G.; Madhusudhan, A.; de Oliveira, R.R. Analysis of Trace Metal Concentrations in Raw Cow's Milk from Three Dairy Farms in North Gondar, Ethiopia: Chemometric Approach. *Environ. Monit. Assess.* **2017**, *189*, 499. [CrossRef] [PubMed]
39. Kodrik, L.; Wagner, L.; Imre, K.; Polyak, K.F.; Besenyei, F.; Husveth, F. The Effect of Highway Traffic on Heavy Metal Content of Cow Milk and Cheese. *Hung. J. Ind. Chem.* **2011**, *39*, 15–19. [CrossRef]
40. Giri, S.; Singh, A.K. Human Health Risk Assessment Due to Metals in Cow's Milk from Singhbhum Copper and Iron Mining Areas, India. *J. Food Sci. Technol.* **2020**, *57*, 1415–1420. [CrossRef]
41. Giri, A.; Bharti, V.K.; Kalia, S.; Kumar, B.; Chaurasia, O.P. Health Risk Assessment of Heavy Metals Through Cow Milk Consumption in Trans-Himalayan High-Altitude Region. *Biol. Trace Elem. Res.* **2021**, *199*, 4572–4581. [CrossRef]
42. Yasotha, A.; Dabadé, D.S.; Singh, V.P.; Sivakumar, T. Risk Assessment of Heavy Metals in Milk from Cows Reared around Industrial Areas in India. *Environ. Geochem. Health* **2021**, *43*, 1799–1815. [CrossRef]
43. Rezaei, M.; Akbaridastjerdi, H.; Jafari, H.; Farahi, A.; Shahabi, A.; Javdani, H.; Teimoory, H.; Yahyaei, M.; Malekirad, A.A. Assessment of Dairy Products Consumed on the Arakmarket as Determined by Heavy Metal Residues. *Health* **2014**, *6*, 323–327. [CrossRef]
44. Shahbazi, Y.; Ahmadi, F.; Fakhari, F. Voltammetric Determination of Pb, Cd, Zn, Cu and Se in Milk and Dairy Products Collected from Iran: An Emphasis on Permissible Limits and Risk Assessment of Exposure to Heavy Metals. *Food Chem.* **2016**, *192*, 1060–1067. [CrossRef]
45. Tizhoosh, M.; Tizhoosh, H. The Concentration of Zinc, Lead, Cadmium and Copper in Raw Milk Production in Industrial Farms in Khorramabad, Iran. *Int. Proc. Chem. Biol. Environ. Eng.* **2016**, *93*, 107–112.
46. Monteverde, V.; Camilleri, G.; Arfuso, F.; Pennisi, M.; Perillo, L.; Patitò, G.; Gioia, G.; Castronovo, C.; Piccione, G. Heavy Metal Levels in Milk and Serum of Dairy Cows from Different Farms Located near an Industrial Area. *Animals* **2022**, *12*, 2574. [CrossRef] [PubMed]
47. Sarsembayeva, N.; Abdigaliyeva, T.; Utepova, Z.; Biltebay, A.; Zhumagulova, S. Heavy Metal Levels in Milk and Fermented Milk Products Produced in the Almaty Region, Kazakhstan. *Vet. World* **2020**, *13*, 609–613. [CrossRef] [PubMed]
48. Khan, N.; Jeong, I.S.; Hwang, I.M.; Kim, J.S.; Choi, S.H.; Nho, E.Y.; Choi, J.Y.; Park, K.S.; Kim, K.S. Analysis of Minor and Trace Elements in Milk and Yogurts by Inductively Coupled Plasma-Mass Spectrometry (ICP-MS). *Food Chem.* **2014**, *147*, 220–224. [CrossRef] [PubMed]

49. Dragusha, B.; Aliu, H.; Dizman, S. Concentrations of Metal Residues in Domestically Produced and Imported Milk in Kosovo. *S. Afr. J. Anim. Sci.* **2022**, *51*, 622–627. [CrossRef]
50. Castro-González, N.P.; Calderón-Sánchez, F.; Fuentes de María-Torres, M.T.; Silva-Morales, S.S.; González-Juárez, F.E. Heavy Metals in Blood, Milk and Cow's Urine Reared in Irrigated Areas with Wastewater. *Heliyon* **2021**, *7*, e06693. [CrossRef]
51. Safonov, V. Assessment of Heavy Metals in Milk Produced by Black-and-White Holstein Cows from Moscow. *Curr. Res. Nutr. Food Sci.* **2020**, *8*, 410–415. [CrossRef]
52. Tahir, M.; Iqbal, M.; Abbas, M.; Tahir, M.A.; Nazir, A.; Iqbal, D.N.; Kanwal, Q.; Hassan, F.; Younas, U. Comparative Study of Heavy Metals Distribution in Soil, Forage, Blood and Milk. *Acta Ecol. Sin.* **2017**, *37*, 207–212. [CrossRef]
53. Pilarczyk, R.; Wójcik, J.; Czerniak, P.; Sablik, P.; Pilarczyk, B.; Tomza-Marciniak, A. Concentrations of Toxic Heavy Metals and Trace Elements in Raw Milk of Simmental and Holstein-Friesian Cows from Organic Farm. *Environ. Monit. Assess.* **2013**, *185*, 8383–8392. [CrossRef]
54. Sujka, M.; Pankiewicz, U.; Kowalski, R.; Mazurek, A.; Slepecka, K.; Góral, M. Determination of the Content of Pb, Cd, Cu, Zn in Dairy Products from Various Regions of Poland. *Open Chem.* **2019**, *17*, 694–702. [CrossRef]
55. Miclean, M.; Cadar, O.; Levei, E.A.; Roman, R.; Ozunu, A.; Levei, L. Metal (Pb, Cu, Cd, and Zn) Transfer along Food Chain and Health Risk Assessment through Raw Milk Consumption from Free-Range Cows. *Int. J. Environ. Res. Public Health* **2019**, *16*, 4064. [CrossRef]
56. Năstăsescu, V.; Mititelu, M.; Goumenou, M.; Docea, A.O.; Renieri, E.; Udeanu, D.I.; Oprea, E.; Arsene, A.L.; Dinu-Pîrvu, C.E.; Ghica, M. Heavy Metal and Pesticide Levels in Dairy Products: Evaluation of Human Health Risk. *Food Chem. Toxicol.* **2020**, *146*, 111844. [CrossRef] [PubMed]
57. Suturović, Z.; Kravić, S.; Milanović, S.; Đurović, A.; Brezo, T. Determination of Heavy Metals in Milk and Fermented Milk Products by Potentiometric Stripping Analysis with Constant Inverse Current in the Analytical Step. *Food Chem.* **2014**, *155*, 120–125. [CrossRef]
58. Rey-Crespo, F.; Miranda, M.; López-Alonso, M. Essential Trace and Toxic Element Concentrations in Organic and Conventional Milk in NW Spain. *Food Chem. Toxicol.* **2013**, *55*, 513–518. [CrossRef] [PubMed]
59. González-Montaña, J.R.; Senís, E.; Gutiérrez, A.; Prieto, F. Cadmium and Lead in Bovine Milk in the Mining Area of the Caudal River (Spain). *Environ. Monit. Assess.* **2012**, *184*, 4029–4034. [CrossRef] [PubMed]
60. Diyabalanage, S.; Kalpage, M.D.; Mohotti, D.G.; Dissanayake, C.K.K.; Fernando, R.; Frew, R.D.; Chandrajith, R. Comprehensive Assessment of Essential and Potentially Toxic Trace Elements in Bovine Milk and Their Feeds in Different Agro-Climatic Zones of Sri Lanka. *Biol. Trace Elem. Res.* **2021**, *199*, 1377–1388. [CrossRef]
61. Bakircioglu, D.; Topraksever, N.; Yurtsever, S.; Kizildere, M.; Kurtulus, Y.B. Investigation of Macro, Micro and Toxic Element Concentrations of Milk and Fermented Milks Products by Using an Inductively Coupled Plasma Optical Emission Spectrometer, to Improve Food Safety in Turkey. *Microchem. J.* **2018**, *136*, 133–138. [CrossRef]
62. Koyuncu, M.; Alwazeer, D. Determination of Trace Elements, Heavy Metals, and Antimony in Polyethylene Terephthalate–Bottled Local Raw Cow Milk of Iğdır Region in Turkey. *Environ. Monit. Assess.* **2019**, *191*, 666. [CrossRef]
63. Bigucu, E.; Kaptan, B.; Palabiyik, I.; Oksuz, O. The Effect of Environmental Factors on Heavy Metal and Mineral Compositions of Raw Milk and Water Samples. *J. Tekirdag Agric. Fac.* **2016**, *13*, 61–70.
64. Zyambo, G.; Yabe, J.; Muzandu, K.; Mkandawire, E.; Choongo, K.; Kataba, A.; Chawinga, K.; Liazambi, A.; Nakayama, S.; Nakata, H.; et al. Human Health Risk Assessment from Lead Exposure through Consumption of Raw Cow Milk from Free-Range Cattle Reared in the Vicinity of a Lead–Zinc Mine in Kabwe. *Int. J. Environ. Res. Public Health* **2022**, *19*, 4757. [CrossRef]
65. Caggiano, R.; Sabia, S.; D'Emilio, M.; Macchiato, M.; Anastasio, A.; Ragosta, M.; Paino, S. Metal Levels in Fodder, Milk, Dairy Products, and Tissues Sampled in Ovine Farms of Southern Italy. *Environ. Res.* **2005**, *99*, 48–57. [CrossRef]
66. Levin, R.; Zilli Vieira, C.L.; Rosenbaum, M.H.; Bischoff, K.; Mordarski, D.C.; Brown, M.J. The Urban Lead (Pb) Burden in Humans, Animals and the Natural Environment. *Environ. Res.* **2021**, *193*, 110377. [CrossRef] [PubMed]
67. Bou Kheir, R.; Shomar, B.; Greve, M.B.; Greve, M.H. On the Quantitative Relationships between Environmental Parameters and Heavy Metals Pollution in Mediterranean Soils Using GIS Regression-Trees: The Case Study of Lebanon. *J. Geochem. Explor.* **2014**, *147*, 250–259. [CrossRef]
68. Araujo, L.B. Influência Dos Municípios Pernambucanos Na Poluição Da Bacia Do Rio Mundaú Com Esgoto Doméstico e. *Rev. Espec.-Line IPOG* **2014**, *1*, 1–13.
69. Ogundiran, M.B.; Ogundele, D.T.; Afolayan, P.G.; Osibanjo, O. Heavy Metals Levels in Forage Grasses, Leachate and Lactating Cows Reared around Lead Slag Dumpsites in Nigeria. *Int. J. Environ. Res.* **2012**, *6*, 695–702. [CrossRef]
70. Lopez-Alonso, M.; Benedito, J.L.; Miranda, M.; Castillo, C.; Hernández, J.; Shore, R.F. Interactions between Toxic and Essential Trace Metals in Cattle from a Region with Low Levels of Pollution. *Arch. Environ. Contam. Toxicol.* **2002**, *42*, 165–172. [CrossRef]
71. Miranda, M.; López-Alonso, M.; Castillo, C.; Hernández, J.; Benedito, J.L. Effects of Moderate Pollution on Toxic and Trace Metal Levels in Calves from a Polluted Area of Northern Spain. *Environ. Int.* **2005**, *31*, 543–548. [CrossRef]
72. Alonso, M.L.; Montaña, F.P.; Miranda, M.; Castillo, C.; Hernández, J.; Benedito, J.L. Interactions between Toxic (As, Cd, Hg and Pb) and Nutritional Essential (Ca, Co, Cr, Cu, Fe, Mn, Mo, Ni, Se, Zn) Elements in the Tissues of Cattle from NW Spain. *BioMetals* **2004**, *17*, 389–397. [CrossRef]
73. Massart, D.L.; Kaufman, L. Hierarchical Clustering Methods. In *The Interpretation of Analytical Chemical Data by the Use of Cluster Analysis*; John Wiley & Sons Inc: New York, NY, USA, 1983; pp. 75–99.

74. Swarup, D.; Patra, R.C.; Naresh, R.; Kumar, P.; Shekhar, P. Blood Lead Levels in Lactating Cows Reared around Polluted Localities; Transfer of Lead into Milk. *Sci. Total Environ.* **2005**, *347*, 106–110. [CrossRef]
75. Derakhshesh, S.M.; Rahimi, E. Determination of Lead Residue in Raw Cow Milk from Different Regions of Iran by Flameless Atomic Absorption Spectrometry. *Am.-Eurasian J. Toxicol. Sci.* **2012**, *4*, 16–19.

animals

Article

Tissue Specific Distribution and Activation of *Sapindaceae* Toxins in Horses Suffering from Atypical Myopathy

Johannes Sander [1,2,*], Michael Terhardt [1], Nils Janzen [1,2], Benoît Renaud [3], Caroline-Julia Kruse [4], Anne-Christine François [3], Clovis P. Wouters [3], François Boemer [5] and Dominique-Marie Votion [3]

1 Screening-Labor Hannover, 30952 Ronnenberg, Germany; m.terhardt@metabscreen.de (M.T.); n.janzen@metabscreen.de (N.J.)
2 Department of Clinical Chemistry, Hanover Medical School, 30625 Hanover, Germany
3 Department of Functional Sciences, Faculty of Veterinary Medicine, Pharmacology and Toxicology, Fundamental and Applied Research for Animals & Health (FARAH), University of Liège, 4000 Liège, Belgium; benoit.renaud@uliege.be (B.R.); acfrancois@uliege.be (A.-C.F.); clovis.wouters@uliege.be (C.P.W.); dominique.votion@uliege.be (D.-M.V.)
4 Department of Functional Sciences, Physiology and Sport Medicine, Faculty of Veterinary Medicine, Fundamental and Applied Research for Animals & Health (FARAH), University of Liège, 4000 Liège, Belgium; caroline.kruse@uliege.be
5 Biochemical Genetics Laboratory, Human Genetics, CHU Sart Tilman, University of Liège, 4000 Liège, Belgium; f.boemer@chuliege.be
* Correspondence: j.sander@metabscreen.de

Simple Summary: Equids kept at pasture are at risk of being intoxicated by ingesting sycamore maple seeds or seedlings, which contain the two non-proteinogenic amino acids hypoglycin A and methylenecyclopropylglycine. These amino acids are converted into effective toxins by metabolic processes, inducing severe damage to oxidative muscles. Toxic metabolites are known to disrupt the cellular use of important energy sources such as short- and medium-chain fatty acids or branched-chain amino acids. The comparative examination of different tissues from five horses that died from this environmental intoxication named atypical myopathy revealed that the highest concentration of active toxins was found in muscles. In all the tissues analyzed, there was still unmetabolized hypoglycin A, which suggests that inhibiting the conversion of protoxins into toxic metabolites would be a possible therapeutic approach.

Abstract: Equine atypical myopathy is caused by hypoglycin A (HGA) and methylenecyclopropylglycine (MCPrG), the known protoxins of sycamore maple (*Acer pseudoplatanus*). Various tissues from five atypical myopathy cases were analyzed but only HGA was found. Whether deamination of MCPrG has already occurred in the intestine as the first stage of metabolization has not been investigated. Activation of the protoxins to methylenecyclopropylacetyl (MCPA)-CoA and methylenecyclopropylformyl (MCPF)-CoA, respectively, occurred mainly in the skeletal muscles, as evidenced by very high concentrations of MCPA-carnitine and MCPF-carnitine in this tissue. Inhibition of the acyl-CoA dehydrogenases of short- and medium-chain as well as branched-chain fatty acids by the toxins led to a strong increase in the corresponding acylcarnitines, again preferentially in skeletal muscles. An accumulation of the long-chain acylcarnitines beyond the level of the control samples could not be detected in the tissues. As a high amount of HGA was always found unmetabolized in the organs, we speculate that targeting the interruption of further metabolization might be a way to stop the progression of intoxication. Inhibition of the mitochondrial branched-chain amino acid aminotransferase, i.e., the first enzyme responsible for the activation of sycamore maple protoxins, could be a therapeutic approach.

Keywords: equine atypical myopathy; hypoglycin A; methylenecyclopropylacetate; *Sapindaceae* toxin; acylcarnitines

1. Introduction

Equids grazing on European pasture are at risk of lethal intoxication by ingesting seeds and seedlings from the sycamore maple (*Acer pseudoplatanus*) [1–6]. Botanically, maple trees belong to the *Sapindaceae* family. Deaths caused by components of these trees have also been described in zoo animals [7–10]. Intoxications by *Sapindaceae* toxins are also of concern for human medicine [11–14]. The proven toxic constituents are hypoglycin A (methylenecyclopropylalanine, HGA) and methylenecyclopropylglycine (MCPrG). Both also occur as dipeptides conjugated with glutamic acid (hypoglycin B and α-glutamyl MCPrG, respectively) [15]. Both HGA and MCPrG are not toxic by themselves. They are rather to be regarded as protoxins that have to be transformed into toxically active compounds.

The metabolism of HGA and MCPrG is known to duplicate the first two steps occurring with branched-chain amino acids [16,17]. The enzymatic activation takes place after the so-called LAT1 transporter, an ubiquitously expressed transport protein from the system L transporter family, has effected the transfer into the intracellular space. The first reaction there consists of the deamination of HGA or MCPrG by a branched-chain aminotransferase (BCAT, EC 2.6.1.42) [18,19]. Thus, MCP-pyruvate is formed from HGA and MCP-glyoxalate from MCPrG.

Unlike other amino acids, the first step of catabolism of the branched molecules does not take place preferentially in the liver but in the skeletal muscles. This gives rise to the assumption that the deamination of HGA and MCPrG can also take place to a large extent in this tissue. One of the aims of our study was to test this theory.

In a second enzymatic step, the α-keto acids formed by transamination are converted by the branched-chain α-keto acid dehydrogenase complex (BCKDHc; EC 1.2.4.4) into the corresponding acyl-CoA derivatives. The activity of the enzymes varies from tissue to tissue, both inter- and intraspecifically, and can also show large differences within an organ [20–22]. The products of the toxin activation are the CoA esters of methylenecyclopropylacetate (MCPA) and methylenecyclopropylformate (MCPF), respectively. Regarding the tissue-specific enzyme activities, we wanted to examine to what extent an activation of the protoxins shows a tissue-typical pattern.

The clinical presentation of *Sapindaceae* intoxication is characterized by acute onset and high lethality. Death usually occurs within 2 days after the onset of the clinical manifestation. An acute rhabdomyolysis syndrome combined with marked hyperglycemia determines the clinical picture in equids [23,24]. In horses, initial knowledge of the pathophysiological process has improved supportive therapy [25] and refined the decision criteria for euthanasia [26].

Necroscopic examination reveals extensive muscle necrosis, predominantly in the postural and respiratory muscles. The extent and severity of the changes vary considerably, not only between individuals but also between the different muscles of each animal. It is worth noting that in some cases, macroscopic lesions are not observed. Within an affected muscle, the degree of myodegeneration may vary greatly. Histology revealed abundant neutral fat, predominantly in type 1 muscle fibers [27,28]. Electron microscopy shows morphological changes in the mitochondria [27,29]. Westermann et al. found deficiencies of short-chain and medium-chain as well as isovaleryl-CoA dehydrogenases in muscle tissue [29]. The plasma/serum acylcarnitines profile showed increased concentrations of short- and medium-chain carnitine esters [29] but also of long-chain acylcarnitines [26]. A further aim of our investigations was therefore to examine whether the expected tissue-specific activation of the protoxins would also lead to a tissue-specific accumulation of products of the energy metabolism.

In some horses, severe macroscopic and histological changes in the myocardium have been found with the presence of pale areas, accumulation of neutral fat, and/or granular degeneration of myocardium cells, while no lesions were found in others [6,23,27,28]. However, elevated troponin I in the blood [26] and both electrocardiographic and echocardiographic changes were seen in most investigated cases [30]. Inhibition of 2-ketoglutarate and pyruvate dehydrogenases in bovine heart mitochondria [31] and reversible inhibition of long-chain acyl-CoA:carnitine acyltransferase were described [32].

Visual examination of other equine tissues such as the central and peripheral nervous systems or the pancreas did not reveal specific lesions; however, forebrain swelling with cellular oedema was reported as well as some little hemorrhage foci on the meninges in the first clinical series. Apart from hyaline, granular, or myoglobin-containing cylinders in the tubules of the kidney, the parenchymatous organs showed no histological abnormalities [27,28]. In rats, ultrastructural electron microscopic examination revealed swelling of the mitochondria of liver cells and a reduction in matrix density 3–5 h after intraperitoneal injection of a high HGA dose (i.e., 100 mg/kg; [33]).

In the case of poisoning, routine laboratory tests show a severe increase in muscle enzyme activity and, most of the time, hyperglycemia, high haptoglobin concentration, hyperlipaemia, high troponin I, and increased liver enzyme activity [23,24,26]. In the urine of horses, myoglobin is excreted in large quantities [24,29]. Such analyses are valuable diagnostic tools in cases of *Sapindaceae* intoxication. However, they do not provide any information about the affected tissues.

The therapy of maple intoxication has so far only been successful to a very limited extent. If one wishes to improve it decisively, more precise knowledge of the pathobiochemistry is required. This study is intended to make a contribution to this.

2. Materials and Methods

2.1. Horses

Five horses with a tentative diagnosis of atypical myopathy according to a diagnostic algorithm [34] used in the literature [26,35,36] were included in this study (Table 1). Serum samples were obtained from four of them just prior to euthanasia being decided due to the progressively worsening of respiratory difficulties [37]. No blood sample was available for one horse that was found dead in pasture. All five horses were necropsied between 1 and 4 days after spontaneous death or euthanasia.

Table 1. Characteristics of atypical myopathy-affected horses.

Horse	Breed	Sex	Age (in Years)	Date of First Signs	Date of Death	Cause of Death
1	Quarter-horse	Female	>24	unknow	3 November 2020	Found dead
2	Spanish Purebred	Stallion	7	31 October 2020	3 November 2020	Euthanasia
3	Belgian Draft Horse	Stallion	3.5	31 October 2020	1 November 2020	Euthanasia
4	Half-blood	Stallion	1.5	27 October 2020	31 October 2020	Euthanasia
5	Percheron	Gelding	4	24 October 2020	26 October 2020	Euthanasia

Samples of the *semitendinosus, triceps brachii, gluteus medius*, myocardium, and diaphragm muscles were taken, as well as samples of the liver, kidney, and pancreas. The cerebrospinal fluid was collected from three horses during necropsy. In addition, in the four euthanized animals, collection of samples of the *semitendinosus* muscle took place not only at necropsy but also immediately after death.

Tissues from five slaughtered animals served as controls. The pancreas and cerebrospinal fluid of unaffected horses were not available. An undefined number of days elapsed between slaughtering and the acquisition of the material, during which the samples had been kept refrigerated but not frozen. Serum samples for control were taken from material examined for diseases not related to maple intoxication. Once collected, all samples were stored frozen at −18 °C until analysis.

2.2. Preparation and Use of Tissue Extracts

Quantities of 30 to 130 mg of material were separated from the frozen tissue samples using a scalpel, and 1000 µL of methanol was added. After brief mixing, samples were placed in an ultrasonic bath for 40 min at room temperature. Centrifugation at 14,000× *g* for 10 min produced a clear supernatant. A dilution factor was calculated from the ratio of the tissue weight to the amount of methanol. Analyses of liquid materials were conducted on 30 µL each.

It is not possible to give a completely accurate indication of the concentrations of protoxins, toxins, their metabolites, or acyl compounds because it is not possible to load animal tissues with these compounds in a reproducible manner. However, repeated extractions can give an indication of the degree of completeness of the initial extraction. It was found that only about 10% of the original concentration is obtained with a second extraction. Therefore, we assume that our analytical values correspond to about 90% of the total tissue contents.

2.3. Quantification of Toxins, Toxin Metabolites, and Several Acyl Metabolites

Ultra-performance liquid chromatography coupled with tandem mass spectrometry using a Xevo TQMS (Waters, Eschborn, Germany) was used for the quantification of HGA and MCPrG plus their metabolites in serum and tissue extracts, as described in detail earlier [38–41]. In addition, concentrations of C4 to C10 acyl conjugates were determined. This method also allowed for the differentiation of branched and unbranched C4 and C5 metabolites. Further carnitines and glycines were determined comparatively in four tissues of one horse. Statistical data on the normal distribution of concentrations of acyl compounds in the tissues of healthy horses are not available. Therefore, in this study, mean values obtained by examination of tissues from five healthy slaughtered animals were used for comparison.

Analyses were performed after butylation. The butyl esters were detected in electrospray ionization-positive mode by multiple reaction monitoring. For the separation of analytes, 5 µL of the final extracts were injected into an Acquity UPLC BEH C18 1.7 µm, 2.1 × 50 mm column (Waters, Eschborn, Germany). For gradient chromatography, acetonitrile/water eluents modified by 0.1% formic acid and 0.01% trifluoroacetic acid were used.

3. Results

3.1. Toxins and Toxin Metabolites

Neither toxins nor toxin metabolites of *Sapindaceae* were found in any of the control samples. As shown in Table 2, non-metabolized HGA was detected in all tissues as well as in the cerebrospinal fluid and the serum of the patients. Unmetabolized MCPrG (not listed in Table 2) was detected in trace amounts only. Concentrations of HGA differed based on the type of material analyzed. There were also large individual differences, as can be derived from the wide range of measured values. In skeletal muscles, including the diaphragm, HGA concentrations were generally found to be lower than in the heart, liver, kidney, and serum. The concentrations were particularly high in the pancreas. Concentrations here exceeded those of *M. semitendinosus* on the same horse by factors up to 23.8.

The distribution of the carnitine and glycine conjugates of MCPA and MCPF showed a pattern very different from that of HGA. MCPA-carinitine, formed from MCPA-CoA by conjugation with carnitine, as well as MCPF-carnitine, formed from MCPF-CoA, were present in very high concentrations in skeletal muscles. In contrast, in the cardiac muscle, the level of MCPA-carnitine was on average only 1.3% of the simultaneously present HGA. In the parenchymal tissues too, the metabolization of HGA was lower than in the skeletal muscles. Accordingly, the concentration of HGA was markedly higher there. The measured values for MCPA-carnitine in these organs, with a few exceptions in liver and kidney samples, were considerably lower than 10% of those of HGA.

MCPA-CoA and MCPF-CoA were conjugated with carnitine or glycine to quite different extents. In skeletal muscles, conjugation with carnitine was by far predominant for both CoA compounds. Again, this was not true for the cardiac muscle. Comparatively strong conjugation with glycine was found in the liver and was particularly pronounced in kidney tissue, which is known to have a high activity of glycine acyltransferase.

Table 2. Hypoglycin A, its metabolites (MCPA-carnitine and MCPA-glycine), and metabolites of methylenecyclopropylglycine (MCPF-carnitine and MCPF-glycine) in different tissues and cerebral fluid collected at necropsy as well as in serum collected prior to euthanasia. The concentrations are expressed in nmol/kg in tissues and in nmol/L in the fluid samples.

Organ	Concentration	HGA	MCPA-Carnitine	MCPA-Glycine	MCPF-Carnitine	MCPF-Glycine
M. semitendinosus.	Mean Range	496 103–1379	2078 156–3690	35 6.5–47	7710 539–23,416	20 0.8–43
M. triceps brachii.	Mean Range	536 111–1473	1884 86–4108	43 1.3–75	7260 10–16,100	22 0.1–38
M. gluteus medius	Mean Range	528 195–1259	2883 40–7482	70 16–109	9595 78–25,782	35 5.1–80
Diaphragm	Mean Range	606 307–1360	639 5–1431	123 14–318	4824 6.9–19,056	33 3.3–83
Myocardium	Mean Range	1115 346–3527	15 0.7–54	99 13–165	27 2.1–109	33 6.8–47
Liver	Mean Range	1340 606–3578	98 0.8–460	244 39–542	240 2.6–1142	109 14–307
Kidney	Mean Range	775 198–2029	71 3.0–208	1655 96–3231	596 3.3–2495	309 35–629
Pancreas	Mean Range	2209 184–5237	75 1.6–162	70 21–145	246 14–795	33 21–55
Serum	Mean Range	1831 909–2948	50 36–64	253 154–296	716 541–958	91 54–114
Cerebral fluid	Mean Range	269 213–356	42 2–61	50 <1–67	398 26–503	23 <1–26

Abbreviations: HGA—Hypoglycin A; MCPA—methylenecyclopropylacetyl; MCPF—methylenecyclopropylformyl.

Results of a study on the post-mortem stability of HGA and MCPA-carnitine as well as three acylcarnitines in *M. semitendinosus* are summarized in Table 3. There was only a slight decrease in the concentration of HGA but a significant reduction in the concentration of MCPA-carnitine. For the acylcarnitines examined, there was a decrease in concentration by factors of up to 6.3.

Table 3. Reduction factor for concentrations measured in samples of *M. semitendinosus* collected immediately after death and collected at necropsy.

Horse	Days between Death and Necropsy	HGA	MCPA-Carnitine	Butyryl-Carnitine	Hexanoyl-Carnitine	Octanoyl-Carnitine
2	2	1.1	2.9	2.6	5.1	6.3
3	2	1.2	2.1	1.9	2.1	3.5
4	3	1.2	2.6	1.2	1.9	2.1
5	1	1.1	2.3	1.9	4.4	4.0

Abbreviations: HGA—hypoglycin A; MCPA—methylenecyclopropylacetyl.

3.2. *Acylcarnitines and Acylglycines*

As with toxin metabolites, tissue-specific accumulations of acyl compounds have been observed. Again, the skeletal muscles were particularly affected, as shown in Table 4.

In the four serum samples collected from 4 of the 5 horses with atypical myopathy, an increase in all acylcarnitines and hexanoylglycine was observed compared to samples collected from horses not affected by sycamore maple intoxication. In the tissues, however, this was only partly the case. Severely increased values were found for short- to medium-chain compounds in the skeletal muscles, but with large individual differences. The very high values for butyrylcarnitine prove a strong inhibition of the enzyme short-chain acyl-CoA dehydrogenase (EC 1.3.99.2). As an indication of the inhibition of the ß-oxidation of medium-chain fatty acids and the degradation of branched amino acids, the corresponding carnitine conjugates were found in the skeletal muscles of the diseased horses in considerably elevated concentrations. Additionally, the simultaneous accumulation of the monounsaturated decenoylcarnitine (C10:1), which is typical for cases of medium-chain acyl-CoA dehydrogenase (EC 1.3.8.7) deficiency in humans, was also observed here.

Table 4. Tissue-specific concentrations of straight-chain and branched-chain acyl compounds (in μmol/kg) in (**A**) atypical myopathy-affected horses and (**B**) controls.

(A)									
Organ	**Concentration**	**C4-C**	**C6-C**	**C6-G**	**C8-C**	**C10:1-C**	**Isobutyryl-C**	**Isovaleryl-C**	**2-MB-C**
M. semitendinosus.	Mean	1464	66	0.6	14	2.2	49	515	225
	Range	157–3548	13–201	0.1–0.9	2.9–42	0.2–6.3	22–92	12–1010	16–430
M. triceps brachii.	Mean	1232	68	0.7	13	2	52	401	166
	Range	1.8–2241	8.9–130	0.5–1.2	1.1–33	0.2–4.1	<0.1–114	22–703	47–450
M. gluteus medius	Mean	1479	70	1.1	18	6.8	45	489	233
	Range	38–4369	7.3–169	0.2–2.1	0.6–55	<0.1–28	0.2–123	19–1103	22–430
Diaphragm	Mean	1203	47	1.3	6.2	0.3	23	260	78
	Range	0.7–3920	2.7–107	0.2–2.4	0.1–10	<0.1–0.5	<0.1–65	8.7–653	17–218
Myocardium	Mean	29	0.8	1	0.3	-	0.4	8.9	6.5
	Range	3.1–54	0.1–2.9	0.1–1.8	<0.1–0.5		<0.1–0.8	0.4–18	1.9–9.4
Liver	Mean	139	4.7	2.2	1.4	0.4	1.7	33	7.9
	Range	0.7–680	0.2–21	1.1–4.4	<0.1–5	<0.1–0.6	<0.1–5.3	0.8–130	<0.1–28
Kidney	Mean	119	5.3	6.3	0.6	0.4	4.1	43	13
	Range	1.9–492	0.3–20	0.4–18	0.1–2.1	<0.1–0.5	<0.1–8.1	2.1–123	3–36
Pancreas	Mean	47	4.1	1	1	0.2	1.2	20	6.3
	Range	3–149	0.2–9.3	0.1–1.3	0.1–3	<0.1–0.3	0.3–2.6	3.3–43	0.7–9.8
Serum	Mean	41	3.6	4.4	0.8	0.6	2.9	13	6.2
	Range	27–58	2.8–4.4	2.2–6.1	0.7–1	0.4–0.6	1.2–4.3	9.9–20	3.4–8.3
(**B**) unaffected controls (means of 5 samples each)									
Organ	**Concentration**	**C4-C**	**C6-C**	**C6-G**	**C8-C**	**C10:1-C**	**Isobutyryl-C**	**Isovaleryl-C**	**2-MB-C**
Skeletal muscles	Mean	0.2	0.5	0.1	<0.1	<0.1	<0.1	2.8	11.4
Myocardium	Mean	8.2	0.7	-	0.1	<0.1	0.4	0.4	1
Liver	Mean	0.9	0.4	0.3	<0.1	<0.1	1.7	26	5.3
Kidney	Mean	2.6	2.3	0.1	0.1	<0.1	0.7	0.5	0.6
Serum	Mean	0.2	<0.1	<0.1	<0.1	<0.1	0.9	0.1	0.2

Abbreviations: C4-C—butyryl carnitine; C6-C—hexanoyl carnitine; C6-G—hexanoyl glycine; C10:1-C—decenoyl carnitine; 2-MB-C—2-methylbutyryl carnitine; -C—carnitine.

In the myocardium, accumulation of straight-chain acylcarnitines was only observed in two patients. In four cases, increased concentrations of the branched compounds isovalerylcarnitine and 2-methylbutyrylcarnitine were detected, but the maximum concentrations were often more than one order of magnitude lower than those in the skeletal muscle of the same animals.

In the liver, kidney, and pancreas, control levels were strongly exceeded only in some cases; in numerous samples, the measured values were in the control range.

Looking at the long-chain acylcarnitines, we found an enrichment of myristoyl-, palmitoyl-, and stearylcarnitine as well as of oleylcarnitine in the serum. Although higher concentrations of long-chain acylcarnitines were found in the skeletal muscles than in other tissues, similar levels were also observed in some controls. Thus, a toxin-dependent accumulation of long-chain acylcarnitines in these tissues was not proven.

The conjugation of the acyl residues in the skeletal muscles was predominantly with carnitine, while binding to glycine occurred to a much lesser extent. It is also evident that the ratios in the heart muscle and in the other tissues diverge strongly from this.

Table 5 illustrates the strong predominance of conjugation with carnitine versus glycine not only for butyric acid but also for branched fatty acids additionally quantified in the tissues of one horse.

Table 5. Ratio of acylcarnitines per acylglycines in tissues of different organs of horse 1 sampled at necropsy.

Organ	C4-C /C4-G	Isobutyryl-C /Isobutyryl-G	2-MB-C /2-MB-G	Isovaleryl-C /Isovaleryl-G
M. semitendinosus	2343	9383	20262	4320
Liver	183	144	82	121
Kidney	10	55	17	9
Myocardium	14	40	68	22

Abbreviations: C4-C or -G—butyryl carnitine or glycine; 2-MB-C or -G—2-methylbutyryl carnitine or glycine.

4. Discussion

This paper is the first to show that HGA is distributed unhindered to all organs after intestinal absorption and that the protoxins are activated to a very unequal extent in the different organs to form the toxic products. The tissue-specific, varying accumulation of products of the energy metabolism was also demonstrated here for the first time. For the assessment of the quantitative results presented here, however, certain limitations should be considered.

Some uncertainty results from the fact that control tissues from slaughtered horses had to be used. The blood is always drained when the animal is for human consumption, but this is not the case when the animal is kept for necropsy. Tissue subtypes such as certain muscle fibers or the renal medulla and cortex could not be examined separately. Furthermore, it must be considered that the collection of material at any point in time always generates an instantaneous value at that very point in time. Pre-mortem changes may take place, and post-mortem alterations have been shown in this study. In addition, since horses on pastures ingest sycamore maple materials irregularly, quite different phases of resorption, metabolization, and excretion can be encountered at the time of the pathological-anatomical examination. HGA is known to be rapidly absorbed from the intestine. However, subsequent metabolization then takes place over a longer period [40,41]. There must be an accumulation when the toxin is taken up again in the decay phase.

Despite the uncertainties described, some basic statements can be made: one of the most important observations is that there is an unhindered distribution of HGA to different tissues, corresponding to an unmetabolized passage of the physiologic branched amino acids through the liver. The concentrations of HGA, however, were low in the cerebrospinal fluid. In the three cases examined, levels of only 212 to 356 nmol/L were found. Studies on a specific blood/brain barrier have not yet been published.

The metabolization rates of the protoxins proved to be very different in a tissue-typical manner. The very large differences in the concentrations of MCPA and MCPF derivatives found in the different tissues largely rule out an effective exchange of these metabolites between organs. It is also shown that the accumulation of the metabolites coming from the β-oxidation of fatty acids under the influence of the *Sapindaceae* toxins obeys a pattern typical of each tissue.

So far, little attention has been paid to the fact that sycamore maple materials contain not only HGA but also MCPrG [15,42], which, however, was not identified in significant concentrations in the serum or in the tissues in our investigation. The reason behind this is unknown. Whether instability in the matrix or rapid metabolization could be responsible requires further investigation. If MCPrG is absorbed in bound form, e.g., as a dipeptide with glutamic acid, it would not be detected with the analytics used in this study. In any case, the high concentrations of MCPF derivatives in skeletal muscle and other tissues suggest that MCPrG is a major contributor to the clinical manifestations of sycamore maple poisoning in equids.

In skeletal muscles, the CoA metabolites of HGA and MCPrG were preferentially conjugated with carnitine (Table 2). The use of carnitine in the treatment of HGA poisoning has occasionally been discussed [25,43]. The very high concentration of carnitine esters in the muscles, even without carnitine application, calls for caution. In contrast, in the liver and especially in the kidney, conjugation of MCPA but not MCPF also occurred to a greater extent with glycine. While MCPA carnitine was only released into the blood to a small extent, the glycine conjugate was detectable in the serum in slightly higher concentrations. The opposite was true for MCPF. An explanation of the tissue- and compound-specific differences would require further investigation.

It should be pointed out that at the time of euthanasia or spontaneous death, HGA was found in considerable quantities in the tissues and blood in an unmetabolized state. Large amounts of not yet metabolized HGA must be expected, especially before the clinical signs are approaching their peak. If one wants to intervene early, it is crucial to prevent further activation of HGA by MCPA-CoA. As HGA activation requires the interaction of several proteins, interruption of the function of one of these might be therapeutically useful. The importance of BCAT for the activation of sycamore maple toxins could be of interest because an inhibitor of the enzyme, gabapentin, is already being used in another veterinary context [44]. Furthermore, previous studies had shown that in the stomach and intestine, HGA can still be found in undigested sycamore maple material and is unabsorbed in the intestinal lumen [8,9]. Gastric/intestinal lavage and the application of activated charcoal could prevent further absorption to a certain extent [45].

5. Conclusions

In addition to HGA testing, a profound diagnosis of atypical myopathy additionally requires the quantitative determination of toxin metabolites as well as evidence of the interrupted ß-oxidation of fatty acids and of the disturbed energetic utilization of the branched amino acids. The subsequent severe skeletal muscle damage results in a sharp increase in serum creatine kinase activity, which confirms the rhabdomyolysis syndrome.

In light of the findings of this study, new therapeutic approaches could be considered, in particular the inhibition of the enzymes and transport systems that allow protoxins to become effective toxins. These possibilities could be used both for prevention (the toxins being ubiquitous in certain regions) and for emergency treatment, as the protoxins are still present in the blood and tissues of poisoned animals.

Author Contributions: Conceptualization, J.S. and D.-M.V.; methodology, J.S.; validation, J.S. and D.-M.V.; formal analysis, J.S.; investigation, J.S., M.T., N.J., B.R., C.-J.K. and A.-C.F.; resources, J.S., M.T., N.J., F.B. and D.-M.V.; data curation, J.S., M.T. and N.J.; writing—original draft preparation, J.S.; writing—review and editing, J.S., M.T., N.J., B.R., C.-J.K., A.-C.F., C.P.W., F.B. and D.-M.V.; visualization, J.S., M.T., N.J. and C.P.W.; supervision, J.S. and D.-M.V.; project administration, D.-M.V.; funding acquisition, D.-M.V. All authors have read and agreed to the published version of the manuscript.

Funding: The study was supported by the la Wallonie agriculture SPW (Service public de Wallonie; Belgique).

Institutional Review Board Statement: Not applicable.

Informed Consent Statement: Not applicable.

Data Availability Statement: The data presented in this study are available upon request from the corresponding authors.

Conflicts of Interest: The authors declare no conflict of interest.

References

1. Baise, E.; Habyarimana, J.A.; Amory, H.; Boemer, F.; Douny, C.; Gustin, P.; Marcillaud-Pitel, C.; Patarin, F.; Weber, M.; Votion, D.M. Samaras and seedlings of Acer pseudoplatanus are potential sources of hypoglycin A intoxication in atypical myopathy without necessarily inducing clinical signs. *Equine Vet. J.* **2016**, *48*, 414–417. [CrossRef]
2. Unger, L.; Nicholson, A.; Jewitt, E.M.; Gerber, V.; Hegeman, A.; Sweetman, L.; Valberg, S. Hypoglycin a concentrations in seeds of acer pseudoplatanus trees growing on atypical myopathy-affected and control pastures. *J. Vet. Intern. Med.* **2014**, *28*, 1289–1293. [CrossRef]
3. Votion, D.M.; Habyarimana, J.A.; Scippo, M.L.; Richard, E.A.; Marcillaud-Pitel, C.; Erpicum, M.; Gustin, P. Potential new sources of hypoglycin A poisoning for equids kept at pasture in spring: A field pilot study. *Vet. Rec.* **2019**, *184*, 740. [CrossRef]

4. Votion, D.M.; Francois, A.C.; Kruse, C.; Renaud, B.; Farinelle, A.; Bouquieaux, M.C.; Marcillaud-Pitel, C.; Gustin, P. Answers to the Frequently Asked Questions Regarding Horse Feeding and Management Practices to Reduce the Risk of Atypical Myopathy. *Animals* **2020**, *10*, 365. [CrossRef]
5. Westermann, C.M.; van Leeuwen, R.; van Raamsdonk, L.W.; Mol, H.G. Hypoglycin A Concentrations in Maple Tree Species in the Netherlands and the Occurrence of Atypical Myopathy in Horses. *J. Vet. Intern. Med.* **2016**, *30*, 880–884. [CrossRef]
6. Zuraw, A.; Dietert, K.; Kuhnel, S.; Sander, J.; Klopfleisch, R. Equine atypical myopathy caused by hypoglycin A intoxication associated with ingestion of sycamore maple tree seeds. *Equine Vet. J.* **2016**, *48*, 418–421. [CrossRef]
7. Renaud, B.; Kruse, C.J.; François, A.C.; Grund, L.; Bunert, C.; Brisson, L.; Boemer, F.; Gault, G.; Ghislain, B.; Petitjean, T.; et al. Acer pseudoplatanus: A Potential Risk of Poisoning for Several Herbivore Species. *Toxins* **2022**, *14*, 512. [CrossRef]
8. Bochnia, M.; Ziemssen, E.; Sander, J.; Stief, B.; Zeyner, A. Methylenecyclopropylglycine and hypoglycin A intoxication in three Pere David's Deers (*Elaphurus davidianus*) with atypical myopathy. *Vet. Med. Sci.* **2021**, *7*, 998–1005. [CrossRef]
9. Hirz, M.; Gregersen, H.A.; Sander, J.; Votion, D.M.; Schanzer, A.; Kohler, K.; Herden, C. Atypical myopathy in 2 Bactrian camels. *J. Vet. Diagn. Investig.* **2021**, *33*, 961–965. [CrossRef]
10. Bunert, C.; Langer, S.; Votion, D.M.; Boemer, F.; Muller, A.; Ternes, K.; Liesegang, A. Atypical myopathy in Pere David's deer (*Elaphurus davidianus*) associated with ingestion of hypoglycin A. *J. Anim. Sci.* **2018**, *96*, 3537–3547. [CrossRef]
11. John, T.; Das, M. Acute encephalitis syndrome in children in Muzaffarpur: Hypothesis of toxic origin. *Curr. Sci.* **2014**, *106*, 1184–1185.
12. Joskow, R.; Belson, M.; Vesper, H.; Backer, L.; Rubin, C. Ackee fruit poisoning: An outbreak investigation in Haiti 2000–2001, and review of the literature. *Clin. Toxicol.* **2006**, *44*, 267–273. [CrossRef]
13. Shah, A.B.; John, T.J. Recurrent Outbreaks of Hypoglycaemic Encephalopathy in Muzaffarpur, Bihar. *Curr. Sci.* **2014**, *107*, 570–571.
14. Shrivastava, A.; Kumar, A.; Thomas, J.D.; Laserson, K.F.; Bhushan, G.; Carter, M.D.; Chhabra, M.; Mittal, V.; Khare, S.; Sejvar, J.J.; et al. Association of acute toxic encephalopathy with litchi consumption in an outbreak in Muzaffarpur, India, 2014: A case-control study. *Lancet Glob. Health* **2017**, *5*, e458–e466. [CrossRef]
15. Fowden, L.; Pratt, H.M. Cyclopropylamino acids of the genus *Acer*: Distribution and biosynthesis. *Phytochemistry* **1973**, *12*, 1677–1681. [CrossRef]
16. Melde, K.; Buettner, H.; Boschert, W.; Wolf, H.P.; Ghisla, S. Mechanism of hypoglycaemic action of methylenecyclopropylglycine. *Biochem. J.* **1989**, *259*, 921–924. [CrossRef]
17. Melde, K.; Jackson, S.; Bartlett, K.; Sherratt, H.S.; Ghisla, S. Metabolic consequences of methylenecyclopropylglycine poisoning in rats. *Biochem. J.* **1991**, *274 Pt 2*, 395–400. [CrossRef]
18. Ghisla, S.; Melde, K.; Zeller, H.D.; Boschert, W. Mechanisms of enzyme inhibition by hypoglycin, methylenecyclopropylglycine and their metabolites. *Prog. Clin. Biol. Res.* **1990**, *321*, 185–192.
19. Osmundsen, H.; Sherratt, H.S. A novel mechanism for inhibition of beta-oxidation by methylenecyclopropylacetyl-CoA, a metabolite of hypoglycin. *FEBS Lett.* **1975**, *55*, 38–41. [CrossRef]
20. Adeva-Andany, M.M.; López-Maside, L.; Donapetry-García, C.; Fernández-Fernández, C.; Sixto-Leal, C. Enzymes involved in branched-chain amino acid metabolism in humans. *Amino Acids* **2017**, *49*, 1005–1028. [CrossRef]
21. Holecek, M. Branched-chain amino acids in health and disease: Metabolism, alterations in blood plasma, and as supplements. *Nutr. Metab.* **2018**, *15*, 33. [CrossRef]
22. Mann, G.; Mora, S.; Madu, G.; Adegoke, O.A.J. Branched-chain Amino Acids: Catabolism in Skeletal Muscle and Implications for Muscle and Whole-body Metabolism. *Front. Physiol.* **2021**, *12*, 702826. [CrossRef]
23. Brandt, K.; Hinrichs, U.; Glitz, F.; Landes, E.; Schulze, C.; Deegen, E.; Pohlenz, J.; Coenen, M. Atypische Myoglobinurie der Weidepferde. *Pferdeheilkunde* **1997**, *13*, 27–34. [CrossRef]
24. Votion, D.-M.; Linden, A.; Saegerman, C.; Engels, P.; Erpicum, M.; Thiry, E.; Delguste, C.; Rouxhet, S.; Demoulin, V.; Navet, R.; et al. History and clinical features of atypical myopathy in horses in Belgium (2000–2005). *J. Vet. Intern. Med.* **2007**, *21*, 1380–1391. [CrossRef]
25. Fabius, L.S.; Westermann, C.M. Evidence-based therapy for atypical myopathy in horses. *Equine Vet. Educ.* **2018**, *30*, 616–622. [CrossRef]
26. Boemer, F.; Detilleux, J.; Cello, C.; Amory, H.; Marcillaud-Pitel, C.; Richard, E.; van Galen, G.; van Loon, G.; Lefere, L.; Votion, D.M. Acylcarnitines profile best predicts survival in horses with atypical myopathy. *PLoS ONE* **2017**, *12*, e0182761. [CrossRef]
27. Cassart, D.; Baise, E.; Cherel, Y.; Delguste, C.; Antoine, N.; Votion, D.; Amory, H.; Rollin, F.; Linden, A.; Coignoul, F.; et al. Morphological alterations in oxidative muscles and mitochondrial structure associated with equine atypical myopathy. *Equine Vet. J.* **2007**, *39*, 26–32. [CrossRef]
28. Palencia, P.; Rivero, J.L. Atypical myopathy in two grazing horses in northern Spain. *Vet. Rec.* **2007**, *161*, 346–348. [CrossRef]
29. Westermann, C.M.; de Sain-van der Velden, M.G.; van der Kolk, J.H.; Berger, R.; Wijnberg, I.D.; Koeman, J.P.; Wanders, R.J.; Lenstra, J.A.; Testerink, N.; Vaandrager, A.B.; et al. Equine biochemical multiple acyl-CoA dehydrogenase deficiency (MADD) as a cause of rhabdomyolysis. *Mol. Genet. Metab.* **2007**, *91*, 362–369. [CrossRef]
30. Verheyen, T.; Decloedt, A.; De Clercq, D.; van Loon, G. Cardiac changes in horses with atypical myopathy. *J. Vet. Intern. Med.* **2012**, *26*, 1019–1026. [CrossRef]
31. Jackson, R.H.; Singer, T.P. Inactivation of the 2-ketoglutarate and pyruvate dehydrogenase complexes of beef heart by branched chain keto acids. *J. Biol. Chem.* **1983**, *258*, 1857–1865. [CrossRef]

32. Entman, M.; Bressler, R. The mechanism of action of hypoglycin on long-chain fatty acid oxidation. *Mol. Pharmacol.* **1967**, *3*, 333–340.
33. Brooks, S.E.; Audretsch, J.J. Studies on hypoglycin toxicity in rats. I. Changes in hepatic ultrastructure. *Am. J. Pathol.* **1970**, *59*, 161–180.
34. van Galen, G.; Marcillaud Pitel, C.; Saegerman, C.; Patarin, F.; Amory, H.; Baily, J.D.; Cassart, D.; Gerber, V.; Hahn, C.; Harris, P.; et al. European outbreaks of atypical myopathy in grazing equids (2006–2009): Spatiotemporal distribution, history and clinical features. *Equine Vet. J.* **2012**, *44*, 614–620. [CrossRef]
35. Gonzalez-Medina, S.; Ireland, J.L.; Piercy, R.J.; Newton, J.R.; Votion, D.M. Equine Atypical Myopathy in the UK: Epidemiological characteristics of cases reported from 2011 to 2015 and factors associated with survival. *Equine Vet. J.* **2017**, *49*, 746–752. [CrossRef]
36. Wimmer-Scherr, C.; Taminiau, B.; Renaud, B.; van Loon, G.; Palmers, K.; Votion, D.; Amory, H.; Daube, G.; Cesarini, C. Comparison of Fecal Microbiota of Horses Suffering from Atypical Myopathy and Healthy Co-Grazers. *Animals* **2021**, *11*, 506. [CrossRef]
37. van Galen, G.; Saegerman, C.; Marcillaud Pitel, C.; Patarin, F.; Amory, H.; Baily, J.D.; Cassart, D.; Gerber, V.; Hahn, C.; Harris, P.; et al. European outbreaks of atypical myopathy in grazing horses (2006-2009): Determination of indicators for risk and prognostic factors. *Equine Vet. J.* **2012**, *44*, 621–625. [CrossRef]
38. Sander, J.; Cavalleri, J.M.; Terhardt, M.; Bochnia, M.; Zeyner, A.; Zuraw, A.; Sander, S.; Peter, M.; Janzen, N. Rapid diagnosis of hypoglycin A intoxication in atypical myopathy of horses. *J. Vet. Diagn. Investig.* **2016**, *28*, 98–104. [CrossRef]
39. Sander, J.; Terhardt, M.; Sander, S.; Aboling, S.; Janzen, N. A new method for quantifying causative and diagnostic markers of methylenecyclopropylglycine poisoning. *Toxicol. Rep.* **2019**, *6*, 803–808. [CrossRef]
40. Sander, J.; Terhardt, M.; Sander, S.; Janzen, N. Quantification of Methylenecyclopropyl Compounds and Acyl Conjugates by UPLC-MS/MS in the Study of the Biochemical Effects of the Ingestion of Canned Ackee (*Blighia sapida*) and Lychee (*Litchi chinensis*). *J. Agric. Food Chem.* **2017**, *65*, 2603–2608. [CrossRef]
41. Sander, J.; Terhardt, M.; Janzen, N. Study on the Metabolic Effects of Repeated Consumption of Canned Ackee. *J. Agric. Food Chem.* **2020**, *68*, 14603–14609. [CrossRef]
42. El-Khatib, A.H.; Engel, A.M.; Weigel, S. Co-Occurrence of Hypoglycin A and Hypoglycin B in Sycamore and Box Elder Maple Proved by LC-MS/MS and LC-HR-MS. *Toxins* **2022**, *14*, 608. [CrossRef]
43. Lieu, Y.K.; Hsu, B.Y.; Price, W.A.; Corkey, B.E.; Stanley, C.A. Carnitine effects on coenzyme A profiles in rat liver with hypoglycin inhibition of multiple dehydrogenases. *Am. J. Physiol.* **1997**, *272*, E359–E366. [CrossRef]
44. Gold, J.R.; Grubb, T.L.; Cox, S.; Malavasi, L.; Villarino, N.L. Pharmacokinetics and pharmacodynamics of repeat dosing of gabapentin in adult horses. *J. Vet. Intern. Med.* **2022**, *36*, 792–797. [CrossRef]
45. Krageloh, T.; Cavalleri, J.M.V.; Ziegler, J.; Sander, J.; Terhardt, M.; Breves, G.; Cehak, A. Identification of hypoglycin A binding adsorbents as potential preventive measures in co-grazers of atypical myopathy affected horses. *Equine Vet. J.* **2018**, *50*, 220–227. [CrossRef]

MDPI AG
Grosspeteranlage 5
4052 Basel
Switzerland
Tel.: +41 61 683 77 34

Animals Editorial Office
E-mail: animals@mdpi.com
www.mdpi.com/journal/animals